Tasks for vegetation science 25

Series Editors

MUT LIETH

sity of Osnabrück, F.R.G.

HAROLD A. MOONEY

Stanford University, Stanford, Calif., U.S.A.

Wetland Ecology and Management: Case Studies

HEL

Unive

Wetland Ecology and Management: Case Studies

Edited by

D.F. Whigham
R.E. Good
J. Kvet

Technical editor

N.F. Good

KLUWER ACADEMIC PUBLISHERS
DORDRECHT / BOSTON / LONDON

Library of Congress Cataloging-in-Publication Data

Wetland ecology and management : case studies / edited by D.F. Whigham, R.E. Good, and J. Kvet ; N.F. Good, technical editor.
p. cm. -- (Tasks for vegetation science ; 25)
Papers presented at the 2nd International Wetlands Conference, in Trebon Czechoslovakia, June, 1984.
Includes index.
ISBN 0-7923-0893-X (alk. paper)
1. Wetland ecology--Congresses. 2. Wetlands--Management--Congresses. I. Whigham, Dennis F. II. Good, Ralph E. III. Květ, J. (Jan), 1933- . IV. International Wetlands Conference (2nd : 1984 : Třeboň, Czechoslovakia) V. Series.
QH541.5.M3W464 1990
333.91'8--dc20 90-43687

ISBN 0-7923-0893-X

Published by Kluwer Academic Publishers,
P.O. Box 17, 3300 AA Dordrecht, The Netherlands.

Kluwer Academic Publishers incorporates
the publishing programmes of
D. Reidel, Martinus Nijhoff, Dr W. Junk and MTP Press.

Sold and distributed in the U.S.A. and Canada
by Kluwer Academic Publishers,
101 Philip Drive, Norwell, MA 02061, U.S.A.

In all other countries, sold and distributed
by Kluwer Academic Publishers Group,
P.O. Box 322, 3300 AH Dordrecht, The Netherlands.

Printed on acid-free paper

Printed in the Netherlands

Contents

INTRODUCTION

This book contains papers on the topics of wetland ecology and management, most of which were presented at the 2nd International Wetlands Conference in Trebon, Czechoslovakia (13-22 June 1984). The conference, hosted by the Hydrobotany Department of the Institute of Botany, was organized by the Czechoslovak Academy of Sciences and the International Wetlands Working Group of the International Association of Ecology (INTECOL) with cooperation from the SCOPE (Scientific Committee on Problems of the Environment) Working Group on Ecosystem Dynamics in Freshwater Wetlands and Shallow Water Bodies, UNESCO Man and the Biosphere (MAB) Program, International Society for Ecological Modelling, and the International Society for Limnology (SIL). Partial sponsorship for the conference and these proceedings was provided by UNESCO (Contract SC/RP/204.079.4) and the Biotic Systems and Resources Program of the U. S. National Science Foundation.

The 2nd International Wetlands Conference was, in part, held in Trebon because of the leading role that the research group there played in developing the field of wetland ecology. In addition to many articles in scientific journals, the group is perhaps best noted for the volume on Pond Littoral Ecosystems that was one of the first books devoted to the dynamics of wetland ecosystems.

Scientific interest in wetlands has expanded enormously during the past 15-20 years and there has been a number of national and international efforts directed toward a better understanding of their ecology and management. Many of these efforts have been initiated because of growing concern about the loss of wetlands and lack of any suitable management plans. The Trebon meeting brought together wetland scientists from all continents and provided an opportunity to exchange valuable information on a variety of aspects on the ecology and management of wetlands.

This volume contains papers that represent aspects of wetland management. Like most ecological topics, the papers clearly demonstrate that the science of wetland management is not evenly developed around the world. In some areas, wetlands have not even been adequately described and there is little information about the impacts that man is having on them. In other areas, information on wetland ecology and management has developed to the point where regulations and laws provide some protection against development. It is our hope this collection of papers will demonstrate that wetlands provide a variety of uses for man and that a range of approaches will be required to protect their valuable functions.

The editors would like to acknowledge the following individuals who provided wordprocessing expertise: Dennis J. Gemmell, Patricia Eager, Srinarayani R. Iyer, and Krishna M. Murali. Michael Siegel prepared many of the figures and Jay O'Neill assisted with the manuscripts while one of us (DFW) was on sabbatical leave. Financial support was provided through the Smithsonian Institution Research Opportunities Fund, the Smithsonian Environmental Research Center - Dr. David L. Correll, Director and the Center for Coastal and Environmental Studies at Rutgers University - Dr. Norbert P. Psuty, Director.

D. Whigham (Edgewater)
R. Good (Camden)
J. Kvet (Trebon)

A REGIONAL STRATEGY FOR WETLANDS PROTECTION

Charles T. Roman[1]

and

Ralph E. Good

Division of Pinelands Research
Center for Coastal and Environmental Studies
Rutgers - The State University of New Jersey
New Brunswick, NJ 08903

ABSTRACT

Wetlands occupy 35% of the New Jersey Pinelands which is a 445,000 ha ecosystem that was named as the first National Reserve in the USA in 1978, and as a Biosphere Reserve by UNESCO in 1983. Wetland types within the Pinelands vegetation mosaic include cedar, hardwood, and pitch pine lowland swamps and scrub-shrub wetlands. In response to national and state legislation, a regional land use planning and management strategy was adopted for protection of the Pinelands ecosystem while also providing for environmentally compatible development. In this paper we report on the various aspects of the Pinelands Comprehensive Management Plan that contribute to protection of the region's wetlands. Based on these efforts in the Pinelands, a conceptual approach to regional wetland protection is presented.

[1]Current Address:
National Park Service
Coastal Research Center
University of Rhode Island
Narragansett, RI 02882

INTRODUCTION

The New Jersey Pinelands are an interrelated complex of uplands, wetlands and aquatic communities, in a largely undeveloped region within the urban corridor of the northeastern United States (Fig. 1). The need to protect the natural and cultural resources of this unique area from encroaching development was recognized, and in 1978 the 445,000 ha New Jersey Pinelands (hereafter referred to as Pinelands) were designated as a National Reserve by the U.S. Congress. In 1983, the Pinelands attained international recognition as a Biosphere Reserve under the Man and the Biosphere Program of the United Nations Educational, Scientific and Cultural Organization (UNESCO). In response to the federal designation and accompanying State of New Jersey legislation, a regional planning and management strategy was developed by the New Jersey Pinelands Commission (1980). The primary goal of this Comprehensive Management Plan, hereafter referred to as the Plan, is to ensure the long-term preservation of the region's natural and cultural resources, while still providing for environmentally compatible development.

The objective of this paper is to identify the planning and regulatory programs of the Plan which contribute to a strategy for effective protection of wetlands which occupy about 35% of the Pinelands. Using the Pinelands experience as a model, a conceptual approach to the protection of wetlands is presented. More detailed summaries of wetland protection in the Pinelands are provided by Zampella and Roman (1983) and Zampella (1985).

PINELANDS WETLANDS: AN ECOLOGICAL OVERVIEW

The Pinelands, located almost entirely on the outer coastal plain, are characterized by a relatively flat topography dissected by streams

D. F. Whigham et al. (eds.), Wetland Ecology and Management: Case Studies, 1–5.

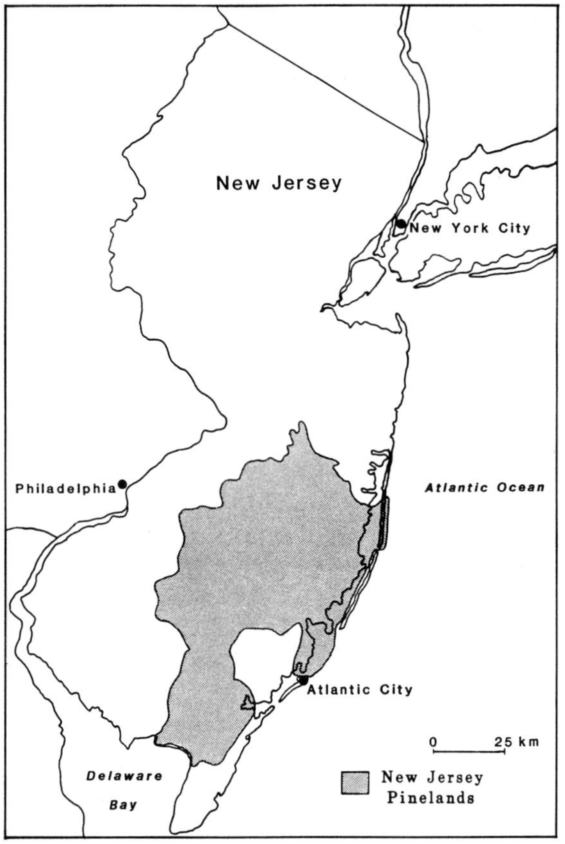

Figure 1. Regional Location of the Pinelands National Reserve.

flowing slowly through a sand-gravel substrate. Stream waters within undeveloped watersheds are acid, pH<4.5, and nutrient poor with nitrate values of <0.05 mg l^{-1} encountered often (Patrick et al. 1979). The palustrine forested wetlands which generally border these streams have been described by Tiner (1985) as being dominated by Atlantic White Cedar (Chamaecyparis thyoides) or several hardwood species (Acer rubrum, Nyssa sylvatica, Magnolia virginiana). Pitch Pine (Pinus rigida) forested wetlands are the most common palustrine type within the Pinelands and often occur in broad areas of poorly drained soils. Shrubs common to the palustrine forested wetland types include Vaccinium corymbosum, Gaylussacia frondosa, and Chamaedaphne calyculata. Forman (1979) provides the most recent review of Pinelands vegetation types and the major factors, such as fire and cultural disturbance, which control the ecological dynamics of Pinelands palustrine inland wetlands, as well as the region's pitch pine and oak (Quercus velutina, Q. alba, Q. prinus) dominated uplands.

Toward the coast, estuarine wetlands are a significant resource of the Pinelands. They are diverse and form a continuum along estuarine bays and streams of Pinelands watersheds. Salt marshes dominated by Spartina alterniflora are especially prevalent, while brackish and freshwater tidal areas may be occupied by Spartina cynosuroides, Scirpus americanus, Typha angustifolia, or Zizania aquatica (Good 1965; Ferren et al. 1981).

THE PINELANDS WETLAND PROTECTION STRATEGY

Comprehensive wetlands protection in the Pinelands is accomplished from both regional and site-specific perspectives. Regionally, the Pinelands acquisition and land allocation programs contribute to the wetland protection strategy, while several environmental performance programs apply protection on a site-specific basis. Further, coordinated efforts between scientists and regulatory agencies provide the research needed to support wetland protection objectives, while education programs inform the general public of the values and functions of wetlands.

Regional Protection

Acquisition represents one of the most effective approaches to the protection of a region's wetlands. In the Pinelands there are over 110,000 ha of publically owned conservation lands (i.e., federal and state wildlife refuges, and state forests). Wetlands, both inland and coastal types, comprise approximately 50% of these environmentally sensitive lands. When stated another way, almost 40% of all Pinelands wetlands are considered as conservation land, and thus, are relatively free from the threat of private development. This insures long-term preservation, especially when considering that the Plan's land acquisition program aims to protect not only wetlands but also entire watersheds or subwatersheds. In addition to government programs, ongoing acquisition efforts by private conservation-oriented organizations contribute to the maintenance and enhancement of Pinelands wetlands from a regional, ecosystem-wide perspective.

A second regional strategy contributing to protection of wetlands is incorporated within the Plan's land allocation program. As reviewed by Good and Good (1984), eight distinct land use management areas were established, each exhibiting varying levels of environmental quality, patterns of existing development, and projected growth needs. One area, the Preservation Area, encompasses over one-third of the Pinelands and represents an area of exceptional environmental quality and is the least populated portion of the Pinelands Reserve. Overall, wetland quality in this core area is high. These wetlands are relatively free from development impacts which can alter water quality, plant and animal species composition and other environmental attributes. Development opportunities in this Preservation Area are extremely limited, thus contributing to the regional protection of wetlands.

The Preservation Area was the only one of the eight land use management areas that was legislatively defined; the remaining management areas were delineated through a detailed analysis of ecological, cultural and other factors with the relative occurrence of wetlands serving as a major analytical factor. For example, Forest Areas (170,060 ha) represent portions of the Pinelands with a high percentage of wetlands, and thus, this land management area exhibits natural resource and cultural characteristics similar to the Preservation Area. Pinelands areas with a lower percentage of wetlands were often assigned to one of the other, less environmentally sensitive, land use areas, such as the Rural Development (58,680 ha) or Regional Growth Area (48,180 ha).

The next important step for each management area was assigning average unit densities for residential development. Densities were assigned according to the total area of privately owned, undeveloped upland which was available (i.e., developable land). In Forest Areas, where wetlands represent a significant natural resource, allowable residential densities average one unit per 6.4 ha of developable land, while the density allocations in Rural Development Areas and Regional Growth Areas are greater. Wetland area is excluded from these density allocation formulas, thereby minimizing the potential for development impacts to affect wetlands on a regional basis. This is especially relevant for the Forest Area. The ultimate outcome of this land allocation program is that development, especially high density development, is directed away from Pinelands areas that have substantial wetland resources. This planning approach minimizes the potential for cumulative impacts to alter regional wetland quality.

Site-specific Protection

The Plan contains several management or regulatory programs each establishing standards that are necessary to insure that permitted development and land use activities proceed with minimal environmental impact. Through the wetlands management program, wetlands and their associated values and functions are afforded priority protection. Foremost to this program is the provision that most development in wetlands is absolutely prohibited. Further, no development can occur within 91 m of a wetland, unless it can be demonstrated that the proposed development will not have a significant adverse impact on the wetland. This requirement represents a critical element to site-specific wetlands protection as a buffer area functions to preserve and maintain a natural upland-to-wetland ecological transition and reduce the potential for impacts associated with adjacent development. Roman and Good (1986) have developed a model to assist the Pinelands Commission in determining appropriate site-specific buffer distances needed to protect wetlands.

The model provides a systematic approach to determine when it would be appropriate to maintain a 91 m buffer and when a lesser buffer would be adequate to protect the wetland. The model is based on a detailed evaluation of wetland quality and on an assessment of potential development-related impacts on wetlands. In addition, the regional land use strategy as set forth in the Plan is incorporated within the buffer delineation model. The model recommends that a buffer of at least 91 m be maintained between wetlands and permitted upland development in the Preservation Area District. For the Forest Area, also an environmentally sensitive area, assigned buffers should range from 91 m to 61 m while for Rural Development and Regional Growth Areas, the recommended buffer distance ranges are 91-30 m and 91-15 m, respectively. This variable buffer provision, or land use factor, facilitates needs for permitting environmentally compatible development in growth-oriented areas, while accentuating the need for priority wetland protection in Pinelands areas exhibiting exceptional ecosystem quality.

Research

The need for effective wetland protection in the Pinelands, and elsewhere nationwide, was based on scientific studies documenting the values and functions of these ecosystems (Good et al. 1978; Greeson et al. 1979; Roman and Good 1983); coupled with the realization that a significant percentage of wetlands have been lost (Frayer et al. 1983). Even with the Pinelands wetland protection strategy now in place, additional scientific research is needed to 1) further support the protection mandates, 2) assess the long-term effectiveness of the wetland protection strategy, and 3) better understand the basic ecological, hydrological and chemical processes of Pinelands wetlands. For example, a recent study of relationships among vegetation, soils and hydrology along upland to wetland transitions will assist Pinelands resource managers in delineating wetland boundaries within the context of the Plan's three-factored wetland definition (Roman et al. 1985). Other studies of palustrine forested wetlands (Ehrenfeld 1983) and aquatic communities (Morgan and Philipp 1986) have compared systems from developed and undeveloped watersheds to evaluate the effects of diffuse source nutrient inputs, hydrologic alterations and other development-related impacts. The results of these studies and others are, in part, used by resource managers to assist with the establishment of environmental standards. With respect to monitoring, a long-term study has recently been initiated to evaluate the effectiveness of 91 m and lesser buffers in protecting wetlands. More basic scientific studies related to Pinelands wetlands have focused on trends in surface water acidity (Morgan 1984), productivity/nutrient dynamics in forested wetlands (Ehrenfeld 1986) and habitat characteristics of the wetland- dependent Pine Barrens tree frog, Hyla andersoni (Freda and Gonzalez 1986; Freda and Dunson 1986). These kinds of research projects are an important element to the maintenance of a scientifically-defensible wetland protection strategy.

Public Education

Environmental education is a most important element to comprehensive and effective wetland protection. The public must be introduced to the ecological and socio-cultural values of wetlands. Upon recognizing that wetlands are a necessary and interrelated part of the biosphere, the public will begin to appreciate the need for strong preservation efforts. The New Jersey Pinelands Commission, other state and federal agencies, conservation-oriented groups and universities have collectively developed an effective public education program. Pamphlets, slide shows, lectures, field trips and school programs all contribute to increasing public awareness and appreciation for Pinelands wetland resources.

CONCEPTUAL WETLAND PROTECTION MODEL

Five important elements for the preservation of wetlands and their associated values and functions are to 1) acquire wetlands within a network of conservation-oriented lands, 2) direct permitted growth away from areas where wetlands are concentrated, 3) prohibit development on and immediately adjacent to wetlands, 4) conduct applied and basic research to define and support wetland protection and management needs, and 5) provide environmental education programs to generate a well-informed public (Fig. 2). Since adoption of the Pinelands Plan in 1980, it has been clearly demonstrated that a wetland protection strategy which employs these elements can be successfully implemented on a regional basis. Comprehensive long-term monitoring, a necessary aspect of the research program, is needed to determine if such a strategy is effectively preserving Pinelands wetlands as ecologically functional ecosystem components.

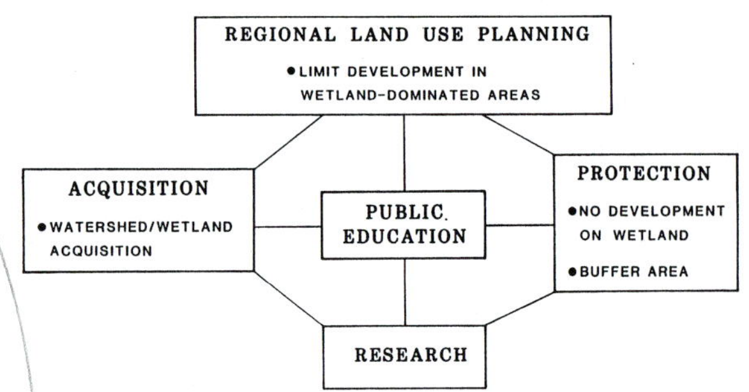

Figure 2. A conceptual strategy for protection of wetlands on a regional basis.

REFERENCES

Ehrenfeld, J.G. 1983. The effects of changes in land-use on swamps of the New Jersey Pine Barrens. Biological Conservation 25: 353-375.

Ehrenfeld, J.G. 1986. Wetlands of the New Jersey Pine Barrens: The role of species composition in community function. American Midland Naturalist 115: 301-313.

Ferren, W.R., Jr., Good, R. E., Walker, R. and Arsenault, J. 1981. Vegetation and flora of Hog Island, a brackish wetland in the Mullica River, New Jersey. Bartonia 48: 1-10.

Freda, J. and Dunson, W. A. 1986. Effects of low pH and other chemical variables on the local distribution of amphibians. Copeia 1986: 454-466.

Freda, J. and Gonzalez, R. J. 1986. Daily movement of the tree frog, Hyla andersoni. Journal of Herpetology 20: 465-467.

Forman, R.T.T., ed. 1979. Pine Barrens: Ecosystem and Landscape. Academic Press, Inc., New York. USA. 601 p.

Frayer, W.E., Monahan, T. J., Bowden, D. C. and Graybill, F. A. 1983. Status and trends of wetlands and deepwater habitats in the conterminous United States, 1950's to 1970's. Department of Forest and Wood Sciences, Colorado State University, Fort

Collins, CO and Office of Biological Sciences, U.S. Fish and Wildlife Service, Washington, DC. USA. 31 p.

Good, R.E. 1965. Salt marsh vegetation, Cape May, New Jersey. Bulletin New Jersey Academy of Science 10: 1-11.

Good, R.E. and Good, N. F. 1984. The Pinelands National Reserve: An ecosystem approach to management. BioScience 34: 169-173.

Good, R.E., Whigham, D. F. and Simpson, R. L. eds. 1978. Freshwater Wetlands: Ecological Processes and Management Potential. Academic Press, Inc., New York. NY. USA. 378 p.

Greeson, P.E., Clark, J. R. and Clark, J. E. eds. 1979. Wetland Functions and Values: The State of Our Understanding. American Water Resources Association, Technical Publication Series, No. TPS79-2. Minneapolis, MN. USA. 674 p.

Morgan, M.D. 1984. Acidification of headwater streams in the New Jersey Pinelands: A re-evaluation. Limnology and Oceanography 29: 1259-1266.

Morgan, M.D. and Philipp, K. R. 1986. The effect of agricultural and residential development on aquatic macrophytes in the New Jersey Pine Barrens. Biological Conservation 35: 143-158.

New Jersey Pinelands Commission. 1980. New Jersey Pinelands Comprehensive Management Plan for the Pinelands National Reserve (National Parks and Recreation Act, 1978) and Pinelands Area (New Jersey Pinelands Protection Act, 1979). New Jersey Pinelands Commission, New Lisbon, NJ. USA. 446 p.

Patrick, R., Matson, B. and Anderson, L. 1979. Streams and lakes in the Pine Barrens. pp. 169-193. In Forman, R. T. T., ed. Pine Barrens: Ecosystem and Landscape. Academic Press, Inc. New York, NY. USA.

Roman, C.T. and Good, R. E. 1983. Wetlands of the New Jersey Pinelands: Values, Functions and Impacts. Division of Pinelands Research, Center for Coastal and Environmental Studies, Rutgers University, New Brunswick, NJ. USA. (revised 1986). 82 p.

Roman, C.T. and Good, R. E. 1986. Delineating wetland buffer protection areas: The New Jersey Pinelands model. pp. 224-230. In Kusler, J. A. and Riexinger, P., eds. Proceedings of the National Wetland Assessment Symposium. Association of State Wetland Managers, Chester, VT. USA.

Roman, C.T., Zampella, R. A. and Jaworski, A. Z. 1985. Wetland boundaries in the New Jersey Pinelands: Ecological relationships and delineation. Water Resources Bulletin 21: 1005-1012.

Tiner, R.W., Jr. 1985. Wetlands of New Jersey. U.S. Fish and Wildlife Service, National Wetlands Inventory. Newton Corner, MA. USA. 117 p.

Zampella, R.A. 1985. Approaches to wetland protection in the New Jersey Pinelands. pp. 375-394. In Kusler, J. and Hamann, R., eds. Wetland Protection: Strengthening the Role of the States. Center for Governmental Responsibility, College of Law, University of Florida, Gainesville, FL. USA.

Zampella, R.A. and Roman, C. T. 1983. Wetlands protection in the New Jersey Pinelands. Wetlands 3: 124-13

SMOLDERING COMBUSTION, THERMAL DECOMPOSITION AND NUTRIENT CONTENT FOLLOWING CONTROLLED BURNING OF TYPHA DOMINATED ORGANIC MAT

Azim U. Mallik
Department of Biology
Lakehead University
Thunder Bay, Ontario
Canada P7B 5E1

ABSTRACT

Blocks of Typha dominated organic mat were subjected to controlled burning in a wind tunnel to study the fire rate of spread, depth of burn and temperature residence time at different drying regimes. C, N, P, K, Ca and Mg contents of the residues were determined following different time-temperature treatments in a muffle furnace. Results of thermal analysis of the mat samples were compared with those of the chemical estimations.

Fire rate of spread and depth of burn were positively correlated with the length of the drying cycles of the mat blocks. In multiple burning, rate of fire spread decreased with increasing number of burns. Smoldering combustion continued with moisture content as high as 387%. Fire temperature in the mat never exceeded 485°C in the wind tunnel. Surface organic matter burned faster with more fluctuations of temperature than that of the lower layer (4 cm below surface). Temperature remained high (400°C) in the lower layer for several hours.

Results of the nutrient analysis showed features common to those found in the thermal analysis. Peaks of the Differential Thermogravimetric (DTG) curves correspond well with the elemental contents of the post-burn mat residues. Nitrogen content increased above 100°C and peaked at 280°C. Around 480°C, P, K, Ca and Mg increased while C and N content declined. K content reached its maximum at 600°C followed by a decline with increasing temperature, P, Ca, and Mg content remained stable beyond 600°C.

Management implications of draining and prescribed burning in the improvement of marsh habitat are discussed.

INTRODUCTION

Revitalization of marsh ecosystems by the introduction of fire is a current trend in managed wetlands (Komarek 1974; Smith 1983; de la Cruz and Hackney 1980). Early studies after naturally occurring fire in marshes (Viosaca 1928; Viosaca 1931; Lynch 1941) and its improvement of wildlife habitat probably provided the stimulus for this kind of management. Burning has resulted in improved game management in wetlands in the United States (Lay 1957; Leege 1969) and there are reports suggesting better primary production and nutritive quality of marsh plants following burning (Linde 1969; Smith and Kadlec 1985; Smith et al. 1984). Changes in plant species composition following draw down and burning have been observed by several investigators, e.g. Millar 1973; van der Valk and Davis 1976; Mallik and Wein 1986.

In many marshes of Atlantic Canada thick floating mats comprised of undecomposed and partially decomposed plant material adversely affect wildlife by restricting the area of open water (Weller 1975) and by reducing species diversity (Grace and Wetzel 1981; Mallik and Wein 1986). The formation and development of floating organic mats and associated vegetation change has been reported in chapter 3 of this volume. It has been suggested that the marsh habitat can be improved by draining and multiple summer burning followed by reflooding. However, although the effects of fire in the marshes have been well documented in terms of population biology of plants, game birds and insects, there is little information about the actual burning process of the marsh, particularly the burning behavior of partially decomposed deep organic matter (Sheshukov 1974; Wein 1983; Stocks 1970).

D. F. Whigham et al. (eds.), Wetland Ecology and Management: Case Studies, 7–17.

The combustion behavior of Typha organic mat was studied under controlled laboratory conditions before applying the technique of draining and summer burning in the field as was suggested by Mallik and Wein (1986). The present chapter examines combustion behaviour of the Typha organic mat including fire temperature, rate of spread of fire, depth of burn, moisture content and bulk density. A study of thermal analysis of Typha organic mat and nutrient content of ash following controlled burning was also undertaken because burning results in thermal decomposition of the substance.

MATERIALS AND METHODS

Fire Behavior in Typha Organic Mat

Five blocks of Typha dominated organic mat each 100 x 35 cm in size were collected in June 1984 from Tintamarre marsh, 8 km northeast of Sackville (lat. 45°56'N, and long. 64°17'W), New Brunswick, Canada. Details of the site description can be obtained from Mallik and Wein (1986). The blocks were transported to the Fire Science Centre, University of New Brunswick and stored at room temperature and relative humidity (RH). Four blocks were burned experimentally one after another in a wind tunnel. Before the first burning each block was placed on the floor of the wind tunnel, the sides of the block were cut smooth and bricks were laid around four sides. The block was then saturated repeatedly with tap water for two days. It was then subjected to drying cycles under 65 lys h^{-1} infrared radiation from the top. The drying cycles before each fire for the first two blocks were 14 hour days and 10 hour nights for two days and for the other two blocks were 16 hour days and 8 hour nights for three days.

Before burning, subsamples were collected for estimation of moisture content and bulk density. Moisture content was measured gravimetrically by oven-drying at 105°C for 48 hours. Bulk density was measured on dry volume basis. During burning, wind speed was maintained at 16 km h^{-1} and temperature between 28 and 30°C with 60% RH. The samples were ignited by a propane torch at the front end of the block facing the wind. When the ignition front was stabilized the rate of fire spread was measured. Temperatures at the surface and 4 cm depth were measured by chromal-alumel thermocouples connected to a chart recorder capable of generating continuous graphs with fire temperature and residence time. Depth of burn following each fire was estimated by placing a pin frame and calculating the difference in surface physiography before and after fire. The first two blocks were subjected to four burns and the second two to three burns.

Thermal Analysis

A preliminary study of thermal decomposition of the Typha mat samples was performed using a small differential thermal analyser (Mettler FP85, Switzerland) which has a maximum temperature of up to 400°C. Small samples (3.5-4.5 g) were weighed in aluminum boats, covered with aluminum lead, and hydraulically pressed. They were then combusted in the thermal analyser.

A more detailed thermal analysis was performed with combustion temperatures up to 1,000°C using a Stanton Redcroft Thermal Analyser STA 781 which simultaneously records thermogravimetric (TG), differential thermogravimetric (DTG) and differential thermal analysis (DTA) curves of the samples. The experiments were carried out in a dynamic (continuous flow) nitrogen atmosphere (30 cm^2 min^{-1}) and at a heating rate of 10°C $min.^{-1}$ The temperature range investigated was 20°C to 1,000°C. The TG curve represents the percentage weight change of the sample, while the DTG curve is its differential form. DTA records the temperature difference between the samples and an inert standard. Thus, the area under a well-defined endothermic peak may be attributed to the energy required to bring about the change. This approach can be used satisfactorily with a single dissociation process (Pope and Judd 1977), namely dissociation of water from soils. Although the quantitative use of DTA to determine absolute energies involved in a process is difficult, a semi-quantitative approach to ascertain relative energies can be obtained by maintaining well characterized standard experimental conditions (Rahman 1982). DTA studies on soils have been reviewed by MacKenzie and Mitchell (1972), who recommended a standard heating rate of 10°C min^{-1} and suggested that conducting the thermal analysis in a nitrogen atmosphere ensures that the soil samples examined represent as near the field conditions as possible.

Eight samples were analyzed following multiple burning of the blocks of Typha organic mat in the wind tunnel. At the end of the fourth burn, ash and charred materials were collected from the surface and then five samples of organic matter were collected from the top towards the bottom at a depth interval of 2 cm.

Nutrient Content in Relation to Burning

Table 1. Rate of fire spread (cm/h^{-1} x ± s.e.) in Typha organic mat in experimental burning in the laboratory. Blocks 1 and 2 were subjected to shorter drying cycles and four burning treatments and blocks 3 and 4 were subjected to longer drying cycles and three burning treatments.

	Rate of Fire Spread (cm h^{-1}) in Burn			
Drying cycles	1	2	3	4
Short (14/10h)	32.0 ± 2.0	31.4 ± 2.0	27.2 ± 9.0	25.7 ± 0.9
Long (16/10h)	60.4 ± 2.2	39.3 ± 0.8	14.2 ± 2.2	-

Temperature

Air dried Typha organic mat samples (0-5 cm deep) were analyzed in duplicate by placing them in a porcelain evaporating dishes and in a pre-heated muffle furnace at 100, 210, 280, 350, 380, 420, 480, 600, 700, 800, 900 and 1,000°C for 30 min. The temperatures were selected on the basis of the endothermic and exothermic peaks of the thermal analysis study and also on the basis of the thermal peaks obtained by other investigators (Fon-Altaba and Montoriol, 1965; Dollimore and Griffiths 1965; Rassonoskaya 1965; Longier-Kuzniarowa 1965; Hoffman and Schnitzer 1965; Schnitzer and Hoffman 1964; Schnitzer et al. 1964; Stewart et al. 1965) working with peat soils. After the temperature treatments total C, N, P, K, Ca and Mg content of the residue were estimated following Jackson (1958).

RESULTS

Rate of Fire Spread and Depth of Burn

The rate of spread of fire in the first burns was higher for the first two blocks than the third and fourth blocks; which were subjected to shorter drying cycles (Table 1). The same trend was maintained in the second and third burns. However, the rate of spread of fire in the third burn of block 2 was more than double compared to the other blocks. In general, the fire rate of spread was decreased with the increasing number of burns (Table 1).

The depth of burn was highest in the first burn in all blocks, and was reduced as the number of burns increased in the blocks subjected to shorter drying cycles. Depth of burn decreased during the second burn in the two blocks subject to longer drying periods but increased again during the third burn (Fig. 1a, b). The depth of burns was very high in the third and fourth blocks (Fig. 1c, d). Following the first deep burn the second burn produced discontinuous fire and the depth of burn was minimum. However, the third burn in the third and fourth blocks was fairly deep. As was expected, the depth of burn was not uniform along the block especially following longer drying cycles (Fig. 2a, c). However, multiple burning following moderate drying produced less variation in the depth of burn along the surface and a more continuous fire (Fig. 1a, b).

Considerable variability existed in depth of burns in close proximities (Fig. 2a, d). It is also evident that areas having higher depth of burn, in one burn generally experienced less depth of burn in the next burn, although there were exceptions, e.g. the left corner of the second block had deep burns both in the first and second burning. This may be due to excessive drying at that corner.

Moisture Content

Moisture content increased from the uppermost part of the mat to deeper areas (Table 2). Following 22 hours of drying the top 2 cm in the first block only had a 12% moisture content. Although there was a general decrease in moisture content at all depths, after burning, the mat was still very wet. The moisture gradient became somewhat irregular as the multiple burning proceeded and the moisture content before and after the fourth burn was fairly low. Smoldering combustion continued with a moisture content of 242%.

In the third and fourth blocks which were given longer drying treatments, up to 14 cm of the top organic matter was removed in the first burn. Moisture contents of these two blocks were appreciably lower than those of the first and second blocks at all comparable depths

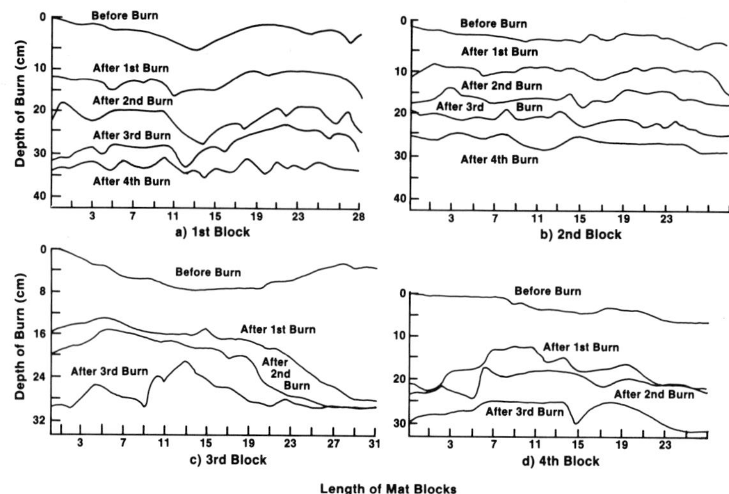

Figures 1a-d. Depth of burn following first, second, thirds, and fourth burn in four blocks of Typha organic mat.

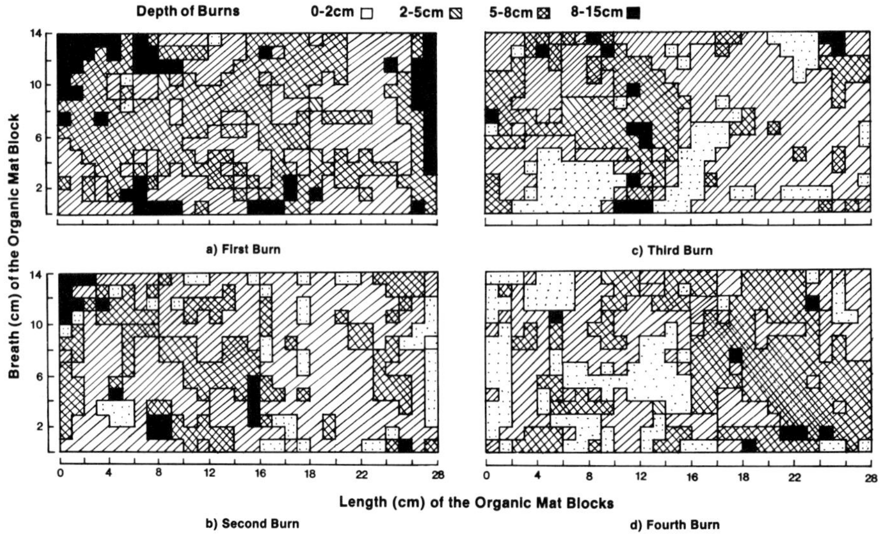

Figures 2a-d. Surface physiography and depth of burn following multiple burning of Typha organic mat in controlled conditions.

(Table 3). In the fourth block smoldering combustion continued with a moisture content of up to 387% (Table 3).

Bulk Density

Bulk density of the organic mat did not vary significantly with depth or between blocks (Tables 4, 5).

Fire Temperature and Residence Time

Fire temperatures never exceeded 485°C. During burning the upper layer of the organic mat burned first with temperatures gradually increasing in the layer underneath (Fig. 3a-d). When the temperature in the upper layer dropped, temperature remained high in the lower layer for several more hours. Sudden temporary fluctuation of surface temperature did not affect the temperature of the lower layer (Fig. 3d).

Following subsidence of the fire, glowing combustion began. This phenomenon is seen as a second peak in temperatures at 4 cm depth in Fig. 3a, c and d. Burning temperatures were

Table 2. Moisture content (percent of dry weight) of the first and second block of organic mat in every 2 cm depth from the surface to the bottom layer before and after multiple burning.

	1st Block					2nd Block		
Depth of soil (cm)	Before 1st burn (saturated)	Before 2nd burn	Before 3rd burn	Before 4th burn	Immediately after 4th burn	Before 2nd burn	Before 3rd burn	Before 4th burn
0-2	265	12	-	-	-	-	-	-
2-4	269	242	-	-	-	-	-	-
4-6	319	364	48	-	-	-	-	-
6-8	556	432	366	10	-	22	-	-
8-10	614	510	492	370	-	100	-	-
10-12	766	549	472	519	169	352	14	-
12-14	770	470	472	523	282	520	138	-
14-16	764	548	456	428	398	620	447	18
16-18	771	614	530	517	352	616	585	169
18-20	900	340	529	306	330	570	529	257
20-22	900	669	335	540	295	307	293	286
22-24	-	-	-	-	-	402	416	407

Table 3. Moisture content (in %) of the third and fourth block of organic mat in every 2 cm depth from the surface to the bottom layer before and after multiple burning.

	3rd Block			4th Block		
Depth of soil (cm)	Before 1st burn	Before 2nd burn	Before 3rd burn	Before 2nd burn	Before 3rd burn	Before 4th burn
0-2	13	-	-	27	-	-
2-4	16	-	-	101	-	-
4-6	24	-	-	249	-	-
6-8	93	-	-	234	-	-
8-10	236	-	-	210	24	-
10-12	220	-	-	191	387	-
12-14	192	-	-	199	379	-
14-16	134	19	-	275	275	-
16-18	84	68	-	243	267	31
18-20	95	162	40	117	255	231
20-22	119	183	213	165	289	228
22-24	230	296	280	-	-	-
24-26	373	357	398	-	-	-

more fluctuating at the surface than underneath (Fig. 3).

Thermal Analysis

Preliminary thermal analysis indicated that there was an overall weight gain from 50°C to 150°C with the maximum at 100°C. This probably happened as a result of endothermic reaction and decomposition of volatile matter and water. From 150°C to 325°C there was a sharp increase in weight loss with the rise of

Table 4. Bulk density (g cm^3)of the first and second block of organic mat in every 2 cm depth from the surface to the bottom layer before and after multiple burning.

	1st Block				2nd Block		
Depth of soil (cm)	Before drying (saturated)	Before 1st burn	Before 2nd burn	Before 3rd burn	Before 2nd burn	Before 3rd burn	Before 4th burn
0-2	.065	.070	-	-	-	-	-
2-4	.111	.110	-	-	-	-	-
4-6	.063	.063	.067	-	-	-	-
6-8	.071	.071	.065	.117	.062	-	-
8-10	.070	.072	.075	.085	.074	-	-
10-12	.066	.066	.075	.086	.056	.167	-
12-14	.061	.061	.102	.112	.056	.142	-
14-16	.076	.080	.095	.125	.067	.108	.184
16-18	.090	.090	.084	.115	.072	.116	.127
18-20	.077	.081	.101	.230	.194	.113	.147
20-22	.091	.102	.220	.101	.193	.197	.136
22-24	-	-	-	-	.114	.150	.107

Table 5. Bulk density (g cm^3)of the third and fourth block of organic mat in every 2 cm depth from the surface to the bottom layer before and after multiple burning.

	3rd Block			4th Block		
Depth of soil (cm)	Before 1st burn	Before 2nd burn	Before 3rd burn	Before 2nd burn	Before 3rd burn	Before 4th burn
0-2	.118	-	-	.067	-	-
2-4	.098	-	-	.053	-	-
4-6	.077	-	-	.093	-	-
6-8	.079	-	-	.099	-	-
8-10	.068	-	-	.073	.083	-
10-12	.078	-	-	.070	.081	-
12-14	.050	-	-	.139	.130	-
14-16	.090	.111	-	.110	.245	-
16-18	.078	.099	-	.170	.216	.243
18-20	.081	.092	.095	.403	.203	.355
20-22	.126	.0978	.085	.286	.202	.331
22-24	.082	.088	.086	-	-	-
24-26	.076	.135	.104	-	-	-

temperature. The curve also showed small peaks at 220, 340 and 390°C and depressions at 210, 240, 350 and 400°C.

The results of the weight loss from samples subjected to detailed thermal analysis are presented in Table 6. Typical real instrumental plots of the thermal analysis of control and 8-10 cm deep samples are presented in Fig. 4 and 5, respectively. The

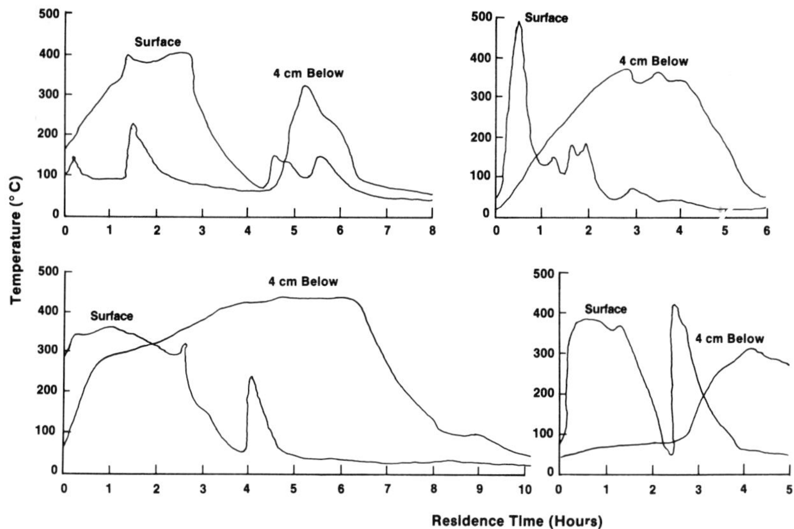

Figures 3a-d. Fire temperature and residence time in Typha organic mat during controlled burning.

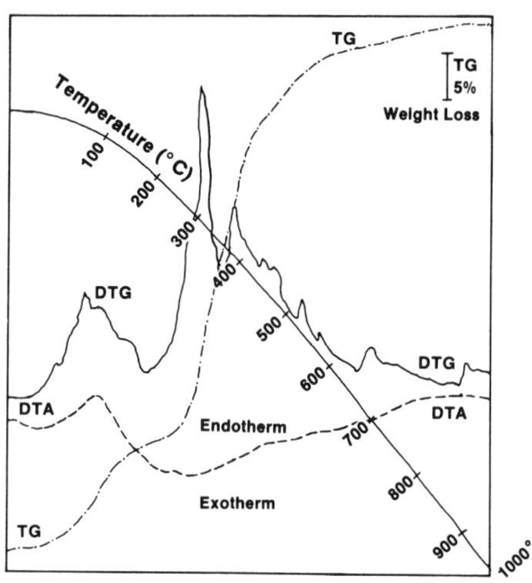

Figure 4. Direct trace of thermal analysis on surface sample of Typha dominated organic mat. Exotherm and endotherm only relevant to DTA. TG, thermogravimetric: DTG, differential thermogravimetric; and DTA, differential thermal analysis curves.

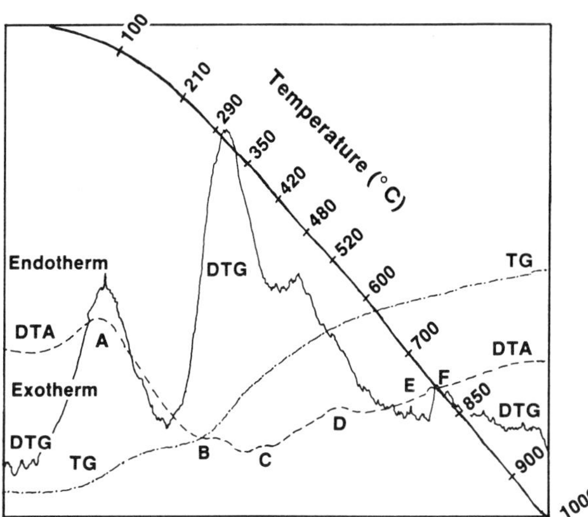

Figure 5. Direct trace of thermal analysis on 8-10 cm Typha organic mat. TG, thermogravimetric, DTG, differential thermogravimetric, and DTA, differential thermal analysis curves.

main features of the thermal analysis results are as follows: (1) TG shows general profile of weight loss with temperature (°C); (2) DTG is the first differential of TG, e.g., it represents W/dT. Where there is a rate of change in weight DTG is represented as a peak.

The main features (which are not clearly shown in TG) are better shown in DTG. This first peak is a broad peak with its maximum at 107°C. This is probably due to dehydration of early removable water. It is a common feature in all the samples, but the magnitude varies. This corresponds approximately to weight gain at 100°C.

The next major peak is characterized by a sharp rise with the maximum around 315 ± 10°C. This could be as a result of decomposition of cellulosic material. There are two other decomposition peaks; one at 470 ± 10°C and the other 760 ± 20°C.

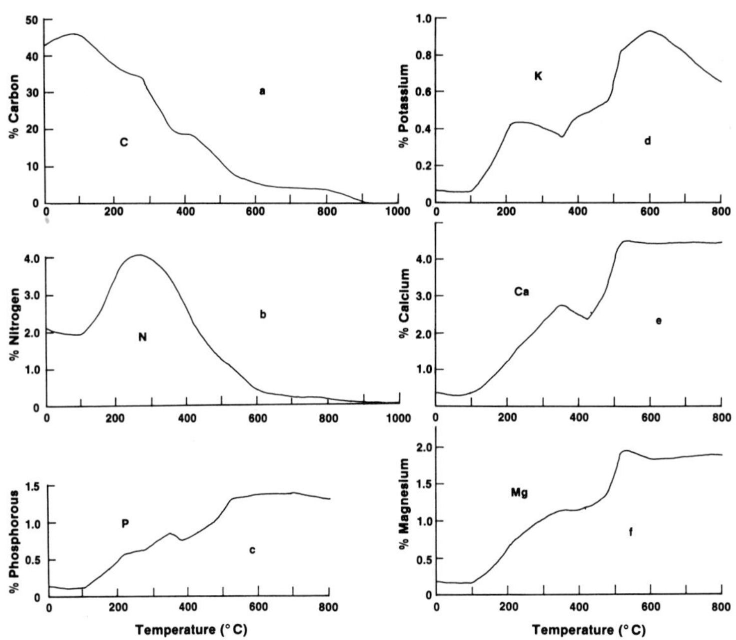

Figures 6a-f. Effect of fire temperature on % carbon, nitrogen, phosphorus, calcium and magnesium content of ash (by dry wt of ash).

Nutrient Analysis in Comparison with Thermal Analysis

Results of the nutrient analysis show features common to those found in the thermal analysis (Fig. 4, 5, and 6). The peak and sharp decline in DTG curves (Fig. 4, 5) around 100°C follows the plots of carbon content closely (Fig. 6a). Nitrogen increased above 100°C and peaked at 280°C (Fig. 6b) which corresponds to the second DTG peak (Fig. 4, 5). Around 480°C phosphorus, potassium, calcium and magnesium contents increased while carbon and nitrogen contents declined (Fig. 6a-f). This was represented by a peak near 480°C in the DTG curves of the thermal analysis (Fig. 4, 5). Potassium concentrations reached a maximum at 600°C followed by a decline with increasing temperature (Fig. 6d) whereas phosphorus, calcium and magnesium remained stable beyond 600°C (Fig. 6c, e, f). The DTG at around 705°C (Fig. 4) can be attributed to a slight increase in the magnesium content at that temperature (Fig. 6f).

DISCUSSION

Depth of Burn and Rate of Fire Spread

The depth of burn is positively related to the number and length of the drying cycles but the first burn was always the deepest and the depth of burn decreased with mat depth in subsequent burnings. Rate of fire spread followed the same trend when the drying cycle was shortest except for the second burn of the third block which was the highest of all the burns. Blocks subjected to the hottest drying cycles had first burns which were very deep and had the highest rate of fire spread. The depth of burn was very small in the second burn but again in the third burn increased. This may be due to the removal of moisture from the mat during the smoldering combustion of the second burn.

Moisture content of the organic mat seems to play the most important role in controlling the smoldering process as the deepest burns occurred in the driest blocks. This pattern does not appear to be related to density of the organic mat as bulk density followed no regular pattern. Increases and decreases in burning temperature and re-ignition of the smoldering process at some points was attributed to the varying drying conditions of the material. Rise and fall of temperature and fire residence time at the surface and below the surface followed a very similar pattern as was reported by McMahon et al. (1980). Once ignited, smoldering can proceed with moisture content as high as 387%. Smoldering is a complicated process involving many physical and chemical controls.

Table 6. Total weight loss (percent of dry weight) during thermal decomposition at different temperatures. Results were derived from nine separate TG runs and samples heated at 15°C min^{-1} under N_2 flow atmosphere. The samples used in the thermal analysis were collected at the end of the fourth burn; the control samples received no previous heat treatment.

	Total Weight Loss (%)											
Sampling depth	Temperature (°C)											
	100	210	280	350	420	480	520	600	700	800	900	1000
Ash	1.8	4.4	4.8	5.0	5.0	5.8	6.2	7.5	9.6	12.3	14.8	15.6
Char	2.7	5.6	7.0	10.8	16.9	20.3	23.2	25.7	29.0	32.2	33.8	35.0
0-2 cm	6.1	7.5	12.4	23.8	32.1	36.6	38.7	40.4	42.5	44.0	45.6	46.5
2-4 cm	3.6	7.4	11.5	20.4	26.1	30.2	31.7	33.6	35.5	37.0	38.1	38.7
4-6 cm	6.0	16.1	19.2	32.0	40.1	45.4	48.0	49.9	52.6	54.8	55.7	56.3
6-8 cm	3.0	5.6	9.0	15.3	19.5	22.2	23.6	25.5	26.9	28.7	30.0	31.1
8-10 cm	3.8	4.9	6.7	10.5	13.1	15.2	16.9	18.1	19.4	20.8	21.7	22.6
Control	9.0	12.25	20.2	36.6	48.0	52.7	54.5	56.0	58.1	59.6	60.75	61.4

Wein (1983) presented a simplified model to explain energy, moisture and gas fluxes to and from the zone of combustion during the burning of an organic layer resting on mineral soil. Interaction of all these factors controls the combustion process. Such detailed study of smoldering process in Typha organic mat has yet to be performed.

Thermal Analysis and Nutrient Analysis

Interpretation of DTA (which represents the energy changes with temperature) is much more complicated. Only the first low temperature endothermic peak at 107°C ("A" in the instrumental plot) is clearly coincidental. This means that it is only a physical removal of some product, probably water. The increasing temperature beyond 230°C brings about decomposition reactions which are chemical and degradative in nature. Hence, every chemical reaction, such as combustion, reverses the sign, so the line ABCDEF is shown as a broad exotherm (Fig. 5). In this region the main weight losses are also reflected in DTA by small humps (endotherms in a broad exotherm) at points B and D, associated with corresponding weight loss peaks (as seen from DTG lines). The small DTA peak at F also corresponds to a high temperature DTG peak.

Clear and quantitative correspondence between DTA and DTG is possible if the forces are purely physical. But as chemical reaction like degradation combustion takes place, it is difficult to analyse them. However, the DTA supports the main features of DTG.

Apart from the comparison of the effectiveness of the two techniques, thermal analysis and estimation of nutrient content, the present paper indicates that the technique of thermal analysis can be used effectively as a method of understanding the thermal degradation/decomposition process in the burning of organic matter.

When the peaks of DTG curves of Figs. 4 and 5 are compared with those of Figs. 6a-f it can be seen that DTG peaks correspond well with the temperatures at which different elements are lost. Between 100 and 200°C, most of the carbon compounds are degraded. From 200 to 300°C maximum nitrogen loss takes place, followed by phosphorus, potassium, calcium and magnesium at temperatures above 500°C. The peaks of DTG are thus good indicators of the thermal degradation of different compounds and show the temperatures at which various elements are lost. Thermal analysis is a much quicker process for demonstrating the effects of burning temperatures on the elemental concentrations of the ash compared to the detailed chemical analysis of the ash following each temperature treatment.

Management Implications

Draining and subsequent multiple burning as a management technique to get rid of floating mat in the marsh is a theoretical idea. Application of multiple burning on a drained marsh is not as simple as it was in the controlled laboratory conditions. As a follow up of the laboratory study presented in this paper some field studies were carried out during the summer of 1984 at the Tintamarre marsh. It was possible to remove the standing dead vegetation and about 3-5 cm surface organic matter by the first burn. However, it was not possible to ignite the burned area for the second or third time except with a very limited success in some small pockets. Incomplete draining and rainy weather made it difficult to re-ignite the material after the first burn. Deep burning by smoldering combustion was only possible where the organic mat was moderately dry but the area that received this kind of combustion was not significant compared to the total area of the marsh. However, in a properly drained marsh following a few days of continuous sunshine the technique is worth repeating and may yield some good results.

In the event of uniform severe smoldering combustion, the regeneration potential of the existing vegetation is impossible due to the total destruction of plant propagules and soil seed bank. In this case regeneration must begin from an external seed source carried by wind and reflooding water which may take a long time depending on the availability of propagules. However, once established, better nutrient conditions of the habitat would enhance vegetation regrowth. In field situations, burning tends to be patchy rather than uniform and thus the possibility of propagule survival is more likely to ensure quicker regeneration following burning. Viable seeds may be found at a depth of 10-12 cm (J.M. Moore, unpublished). Patchiness of vegetation is more desirable from a wildlife point of view.

ACKNOWLEDGEMENT

The research was conducted during the tenure of a Post-doctoral fellowship at the Fire Science Centre, University of New Brunswick, Canada. I thank Dr. Ross W. Wein, for his help and advice during the study. Thanks are also due to Drs. R.J. West and E.H. Hogg for their valuable comments on the manuscript. Help of Drs. A.A. Rahman and H.H. Krause in thermal analyses and nutrient analyses is gratefully acknowledged.

REFERENCES

de la Cruz, A.A. and Hackney C.T. 1980. The effects of winter fire and harvest on the vegetational structure and primary productivity of two tidal marsh communities in Mississippi. Publication M-AS6P-80-013. Mississippi-Alabama Sea Grant Consortium, Ocean Springs, MS. USA.

Dollimore, D. and Griffiths, D.L. 1965. The differential thermal analysis of metal oxalates in controlled atmospheres. pp. 126-127. In Redfern, J.P., ed. Thermal Analysis 1965. MacMillan & Co. Ltd., London.

Fon-Altaba M. and Montoriol-Pous, J. 1965. Differential thermal analysis and thermal treatment of some coloured Spanish calcites. pp. 212-213. In Redfern, J.P., ed. Thermal Analysis 1965.MacMillan & Co. Ltd., London. U.K.

Grace, J.B. and Wetzel, R.G. 1981. Habitat partitioning and competitive displacement in cattails (Typha): experimental field studies. American Naturalist 118: 463-474.

Hoffman, I. and Schnitzer, M. 1965. Thermogravimetric studies on soil humic compounds. pp. 62-63. In Redfern, J.P., ed. Thermal Analyses 1965, MacMillan & Co. Ltd., London. U.K.

Jackson, M.L. 1958. Soil chemical analysis. Prentice Hall Inc., Englewood Cliffs, N.J. USA.

Komarek, E.V. 1974. Effects of fire on temperate forests and related ecosystems: southeastern United States. pp. 251-277. In Kozlowski, T.T. and Ahlgren, C.E., eds. Fire and Ecosystems. Academic Press. New York, NY. USA.

Lay, D.W. 1957. Browse quality and the effects of prescribed burning in southern pine forests. Journal of Forestry 55: 342-347.

Leege, T.A. 1969. Burning seral brush ranges for big game in northern Idaho. Trans. North American Wildlife and Natural Resources Conference 34:429-438.

Linde, A.F. 1969. Techniques for wetland management. Wisconsin Department of Natural Resources Research Report 45, 156 p.

Longier-Kuzniarowa, A. 1965. Standard derivatograms of clays. pp. 196-197. In Redfern, J.P., ed. Thermal Analysis 1965, MacMillan & Co. Ltd., London. U.K.

Lynch, J.J. 1941. The place of burning in management of the Gulf Coast wildlife refuges. Journal of Wildlife Management 5: 454-457.

MacKenzie, R.C. and Mitchell, B.D. 1972. pp. 267-297. In MacKenzie, R.C., ed. Differential Thermal Analysis Vol. 2, Academic Press, London. U.K.

Mallik, A.U. and Wein, R.W. 1986. Response of Typha marsh community to draining, flooding and seasonal burning. Canadian Journal of Botany 64: 1236-2143.

McMahon, C.K., Wade, D.A. and Tsoukalas, S.N. 1980. Combustion characteristics and emissions from burning organic soils. Proceedings 23rd Annual Meeting of the Air Pollution Control Association, Montreal, Quebec, Canada.

Millar, J.B. 1973. Vegetation changes in shallow marsh wetlands under improving moisture regime. Canadian Journal of Botany 51: 1443-1457.

Moore, P.D. and Bellamy, D.J. 1974. Peatlands. Springer,New York. USA. 221 p.

Pope, M.I. and Judd, M.D. 1977. pp. 30-43. In Differential thermal analysis - a guide to techniques and its application. London, Heyden. U.K.

Rahman, A.A. 1982. Mechanism of hydration and the role of an admixture in cement - a thermal analytical approach. pp. 1310-1317. In Miller, B., ed. Thermal Analysis. Vol. II. John Wiley and Sons, NY. USA.

Rassonoskaya. 1965. Differential thermal analysis of salts. pp. 120-121. In Redfern, J.P., ed. Thermal Analysis 1965, Macmillan & Co. Ltd., London. U.K.

Schnitzer, H. and Hoffman, I. 1964. Pyrolysis of soil organic matter. Soil Science Society of America Proceedings 28:520-525.

Schnitzer, M., Turner, R.C. and Hoffman, I. 1964. A thermogravimetric study of organic matter of representative Canadian podzol soils. Canadian Journal of Soil Science 44:7-13.

Sheshukov, M.A. 1974. Features of sward (turf) fires. Lesovedenie, 5:81-83 (in Russian).

Smith, L.M. 1983. Effects of prescribed burning on the ecology of a Utah marsh. Dissertation. Utah State University, Logan, Utah. USA.

Smith, L.M. and Kadlec, J.A. 1985. Fire and herbivory in Great Salt Lake Marsh. Ecology 66(1): 259-265.

Smith, L.M., Kadlec, J.A. and Fonnesbec, P.V. 1984. Effects of prescribed burning on nutritive quality of marsh plants in Utah. Journal of Wildlife Management, 48(1): 285-288.

Stewart, J.M., Birmie, A.C. and Mitchell, B.D. 1965. Thermal analysis of Scottish hill peat. pp. 64-65. In Redfern, J.P., ed. Thermal Analysis 1965, MacMillan & Co., Ltd., London. U.K.

Stock, B.J. 1970. Moisture in the forest floor - its distribution and movement, Canadian Forestry Series Publication, 1271. 20 p.

van der Valk, A.G. and Davis, C.B. 1976. Changes in the composition, structure, and production of plant communities along a perturbed wetland coenocline. Vegetatio 32(2): 87-96.

Viosca, P. Jr. 1928. Louisiana wetlands. Ecology 9: 216-229.

Viosca, P.Jr.1931. Spontaneous combustion in the marshes of southern Louisiana. Ecology 12: 432-442.

Wein, R.W. 1983. Fire behavior and ecological effects in organic terrain. pp. 81-95. In Wein, R.W. and MacLean, D.A., eds. The Role of Fire in Northern Circumpolar Ecosystems. John Wiley & Sons, New York, NY. USA.

Weller, M.W. 1975. Studies of cattail in relation to management for marsh wildlife. Iowa State Journal of Science 49(4): 383-412.

MICROSCALE SUCCESSION AND VEGETATION MANAGEMENT BY FIRE IN A FRESHWATER MARSH OF ATLANTIC CANADA

Azim U. Mallik
Department of Biology
Lakehead University
Thunder Bay, Ontario
Canada P7B 5E1

ABSTRACT

Plant species composition of a Typha-dominated freshwater marsh was strongly influenced by the thickness of the floating organic mat. The buildup of organic matter was negatively correlated with pH (-0.63) and positively correlated (0.85 and 0.80) with the mat thickness and water table depth, respectively. The tabulation of species cover data and the canonical correlation analysis of the environmental and plant factors arranged the 48 quadrats into four groups reflecting the seral stages towards the initial conditions of relatively stable fen formation. The buildup of undecomposed organic matter seems to be the driving force for the acidification, which is further intensified by the presence of Sphagnum and other bog-forming species.

The Typha marsh community was subjected to draining and seasonal burning treatments to control the growth of emergent aquatics. Treatments resulted in an increase in total number of species after three years. Cover and frequency of Aster novi-belgii, Lycopus uniflorus, Epilobium watsonii, Brachythecium salebrosum, Pleurozium schreberi, and Cladonia cristatella increased appreciably on the drained side whereas those of Carex spp., Lysimachia terrestris, Epilobium palustre, Pellia epiphylla, Sphagnum squarrosum, Drepanocladus exannulatus, and Helodium blandowii increased on the flooded side. Draining plus summer burning produced the lowest cover, stem density, plant height, and stem base diameter of Typha. An attempt was made to interpret the effects of disturbance on the natural paludification process that leads to the development of patches of fen within the marsh.

INTRODUCTION

Marshlands, which constitute a significant part of the Maritime provinces, are among the most valuable and irreplaceable environments in Atlantic Canada (Roberts and Robertson 1986). Ganong (1903) published a comprehensive description of early land reclamation practices in the marshes of the Bay of Fundy which at that time occupied an area of 30,000 ha. A general ecological overview of these marshes can be obtained from Gauvin (1979) and Teal and Teal (1969).

To facilitate an understanding of the present ecology of the Bay of Fundy marshes one ought to know the sequence of events in their use by man. Wetlands in this area were formed by the deposition of mineral silts from the Bay of Fundy tides. In general, salinity decreases with distance from the sea, thus forming salt marshes along the coast and freshwater marshes, fens and bogs further inland. Beginning as early as the sixteenth century, tidal wetlands were diked, ditched, and used for pasture and hay production by the Acadian settlers. At various intervals, tidal water was allowed to reflood the land and deposit more silt. Almost all the lakes and bog areas at the head of the Bay of Fundy were drained and reflooded this way so that the area could be used for agriculture. This practice continued until the 1930's when, due to the decline of hay prices and shortage of labor, most diked marshes were abandoned. With the subsequent lack of tillage and the deterioration of interval drainage, organic matter accumulation increased in many of these marshes.

During the 1950's and early 1960's the Federal Government repaired the dikes and water control structures, however, due to poor economic returns, the agriculture initiative was not practical. In the latter part of the 1960's the Canadian Wildlife Service and the province of Nova Scotia acquired some of the wetlands as

D. F. Whigham et al. (eds.), Wetland Ecology and Management: Case Studies, 19–29.

wildlife management areas. Ducks Unlimited (Canada) provided additional funds and expertise for developing impoundments on these marshes.

In the newly impounded marshes, thick mats of organic matter disengaged from the bottom and floated upward to cover the open water surface. The decomposed and partially decomposed organic mats are buoyant largely due to carbon dioxide and methane evolution as a result of anaerobic decomposition (Hogg and Wein 1987). Extensive floating mats make the habitat unsuitable for wildlife, especially for ducks and other birds. Apart from covering the surface water, large quantities of nutrients remained "locked up" (Rowe and Scotter 1973) in the organic mat. Typha glauca and Sparganium eurycarpum (hereafter mentioned as Typha and Sparganium, respectively) dominate the organic mats. In 1981 the Fire Science Center of the University of New Brunswick in collaboration with the Canadian Wildlife Service and Department de Biologié, Université de Moncton, New Brunswick, embarked on a research project in one of the marshes, Tintamarre Marsh, with floating Typha organic mats (Wein and Krusi 1982).

In this paper I examine the relationship between vegetational changes and mat thickness and the effects of draining and seasonal burning on the plant composition and productivity of Typha. Management implications of microscale plant succession and draining and seasonal burning of the marsh are discussed. Nomenclature and author citations for vascular plants follows Scoggan (1978-1979), for bryophytes and fungi follows Ireland (1982) and Hale (1961), respectively. Some of the materials presented in this chapter appear in two journal articles in the Canadian Journal of Botany by Mallik (1989) and Mallik and Wein (1986).

BACKGROUND INFORMATION

Study Area

Tintamarre Marsh is a 100 ha impounded freshwater marsh, located 8 km northeast of Sackville at latitude 45°56' N, longitude 64°17.5' W. The marsh, within the Tintamarre National Wildlife Area, developed following abandonment for hay and pasture production. Typha became the dominant species. To improve drainage, Goose Lake Canal was dug in the mid 1960's under Maritime Marsh Reclamation Act). In 1973, the 100 ha Hog Lake Impoundment was diked on four sides and water control structures were installed. The

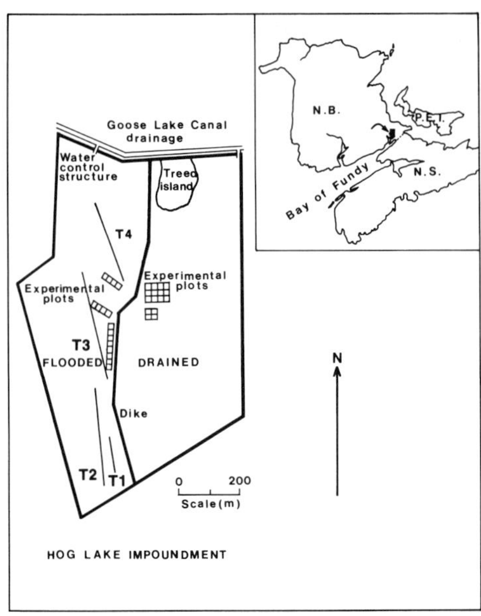

Figure 1. Location of study showing the experimental plots on drained and flooded sides of the Hog Lake impoundment of Tintamarre marsh. N. B., New Brunswick; P. E. I., Prince Edward Island; N.S., Nova Scotia. T1, T2, T3 and T4 are the transects studied (modified from Mallik and Wein 1986, Courtesy of NRC and Canadian Journal of Botany.

impoundment was the site of this study (Fig. 1). In mid-November (1980) water was drained from the impoundment. In December and January, a central dike was constructed to divide the impoundment into two roughly equal-sized areas (Fig. 1). Water control structures have been used to regulate water levels in the two basins. The water control structure on the west basin was closed in February (1981) to catch snowmelt and spring rains. By May, water depths reached the original level in that basin. During the summer of 1981 and 1982 the western basin was maintained at the original level and the eastern basin was drained.

Climate

The climate at the head of the Bay of Fundy is typical of adjacent coastal areas having mean January and July temperatures of -6.9 and 17.9° C, respectively. Of the total mean annual precipitation of 1,044 mm, 851 mm occurs as rain which is fairly evenly distributed throughout the year except for the months of March, April, and July which are somewhat drier. Snowfall usually begins in November and continues until April. The relative humidity in summer months varies from 49 to 62%. In general, the high rainfall and high humidity give the area a low Fire Weather Index (F.W.I.) (Simard 1975). The area experiences fairly uniform, high wind speeds (16 to 20 km h^{-1}) throughout the year. Both these factors are important when considering the use of fire as an ecological tool in the marsh.

Vegetation Characteristics

Following impoundment, the 10-80 cm deep organic mat broke free from the bottom and floated to the surface. In some places, however, partially decomposed organic matter has remained submerged under 5-27 cm of water. On the southern side of the impoundment, the dominant plant is Sparganium. Patches of Typha and aquatic species such as Lemna minor, Ceratophyllum sp., Utricularia intermedia, and Potamogeton foliosus, are also present. Typha is the dominant species on the northern side, with a thick cover of many understory species such as Epilobium leptophyllum, Triadenum virginicum, Potentilla palustris, Viola macloskeyi, Galium palustre, and Carex spp. In places where Typha growth is stunted, a thick cover of Sphagnum fimbriatum and S. squarrosum along with Pseudobryum spp., Pellia epiphylla, Drosera rotundifolia, Carex limosa, and Calamagrostis canadensis developed, giving the marsh a patchy fen/bog element.

Field observations indicated that there was a vegetational gradient from the open water habitat to the thick floating mat of Typha. Those observations lead to the following hypotheses:

i) Typha, with high productivity, is capable of building the organic mats to a substantial thickness and this, together with the resulting buoyancy, strongly determines the vegetation dynamics of the system.

ii) Typha mat-building process creates a very resilient system favorable for Typha growth, but subsequently the habitat conditions gradually change (pH and nutrients decline) resulting in a reduction in Typha dominance.

iii) Sphagnum colonizes Typha mats once they are sufficiently thick to be continuously buoyant. Development of Sphagnum patches results in a further reduction in pH and at this point Typha becomes locally uncommon or extinct and Sphagnum predominates along with other fen and bog species.

iv) Seasonal burning of Typha vegetation on the floating mat or draining plus seasonal burning can be used to remove the Typha and Sphagnum dominance of the marsh.

MATERIALS AND METHODS

Succession Studies

Four transects were established along the gradients from areas with thin mats with vigorous Typha near open water to deep mats with stunted Typha. The transects ranged from 65 to 110 m in length and a total of 48 quadrats (1 m^2) were sampled along them. In each quadrat, percentage cover of each plant species was estimated and a 12.5 cm diameter core was cut from the mat, but outside of the quadrat. The partially decomposed mat core was examined for plant composition and mat thickness and depth to water table on the mat and water depth below the mat were measured in addition to the thickness of the moss (mainly Sphagnum) layer. Water samples were collected from the surface and underneath the mat and were analyzed immediately for temperature and pH.

Data were subjected to a phytosociological analysis (Braun-Blanquet 1932; Wells 1981). Data on mat thickness, Sphagnum thickness, Typha shoot density, total number of vascular plants, water depth above and below the mat, and surface water pH beside each quadrat were also included in the phytosociological analysis.

A canonical correlation analysis (Cooley and Lohnes 1971) was performed on 13 variables: 1) mat thickness, 2) depth to water table on the mat, 3) pH, 4) water depth, 5) total water depth below floating mat, 6) Typha plant height, 7) stem density 8) cover, 9) Sparganium cover, 10) Lemna cover, 11) Sphagnum cover, 12) total number of vascular plants and 13) total number of bryophytes. Data for two Myrica dominated quadrats were excluded from this analysis because an insufficient number of samples were in the group. Since the overall sample size of all quadrats is not large enough relative to the number of variables to be analyzed, Canonical correlation is used in this paper as a descriptive technique to support results from other analysis.

Seasonal Burning

Four replicated blocks, each having four (30 x 30m) plots, were established in the drained and flooded areas in 1981 (Fig. 1). Uniformity of Typha growth was taken as a criterion in selecting the plots. The plots were burned in spring (June 3 and 17), summer (July 1 and 15) and autumn (August 12 and September 16) of 1981.

Three years later, percent cover was visually estimated for all species of vascular plants, bryophytes, fungi and lichens by

Table. 1. Frequency of occurrence and mean (± 1 standard deviation) percent cover for species in relation to the biotic and abiotic factors listed at the bottom of the table.

Species	Group I		Group II		Group III		Group IV	
	Freq.	$\bar{\chi}$ (SD)	Freq.	$\bar{\chi}$ (SD)	Freq.	$\bar{\chi}$ (SD)	Freq.	$\bar{\chi}$ (SD)
Lemna minor	100.0	43.7(27.3)	33.3	11.3(19.2)	-	-	-	-
Sparganium eurycarpum	80.0	47.9(35.8)	27.8	8.0(5.4)	-	-	-	-
Cicuta bulbifera	40.0	1.7(1.2)	16.7	2.0(1.0)	15.4	1.0(0.0)	-	-
Utricularia intermedia	33.3	29.0(24.6)	27.8	23.6(32.0)	-	-	-	-
Riccia fluitans	40.0	12.5(5.2)	-	-	-	-	-	-
Carex pseudo-cyperus	6.7	2.0	66.7	17.2(19.9)	-	-	-	-
Calamagrostis canadensis	13.3	4.0(2.8)	16.7	2.3(1.2)	100.0	16.2(9.3)	100.0	15.0(7.1)
Sphagnum fimbriatum	-	-	33.3	52.2(47.4)	100.0	64.6(33.7)	100.0	11.0(1.4)
Typha glauca	20.0	35.0(5.0)	61.1	31.3(15.2)	100.0	20.6(11.7)	-	-
Viola macloskeyi	-	-	22.2	4.0(4.1)	100.0	16.8(8.8)	100.0	15.0(0)
Lycopus uniflorus	-	-	72.2	4.3(3.0)	92.3	3.4(2.5)	50.0	10
Epilobium leptophyllum	-	-	61.1	4.1(3.2)	92.3	3.0(1.1)	100.0	1.5(0.7)
Pellia epiphylla	6.7	3.0	38.9	30.7(37.2)	55.6	14.0(6.6)	100.0	6.5(4.9)
Carex canescens	-	-	66.7	15.1(18.2)	69.2	12.0(10.3)	100.0	6.0(5.7)
Triadenum virginicum	-	-	72.2	4.2(3.6)	92.3	4.6(2.8)	100.0	6.0(1.4)
Potentilla palustris	-	-	61.1	7.3(4.1)	84.6	4.5(3.3)	-	-
Galium palustre	6.7	1.0	77.8	5.3(3.9)	76.9	1.4(0.5)	100.0	2.5(0.0)
Agrostris hyemalis	6.7	2.0	55.6	6.2(4.7)	46.2	1.8(0.8)	50.0	10.0
Menyanthes trifoliata	-	-	5.6	1.0	69.2	1.9(5.9)	50.0	10.0
Lysimachia sp.	-	-	16.7	2.0(0.0)	61.5	1.9(0.6)	50.0	2.0
Aster novi-belgii	-	-	27.8	1.4(0.5)	61.5	2.8(2.3)	50.0	2.0
Calliergon giganteum	13.3	2.0(0.0)	44.4	8.0(7.0)	38.5	8.4(5.0)	50.0	5.0
Lophocolea heterophylla	-	-	-	-	38.5	4.6(0.9)	-	-
Drepanocladus aduncus	6.7	5.0	38.5	7.6(7.0)	30.8	4.3(3.9)	-	-
Drepanocladus exannulatus	-	-	16.7	35.0(30.0)	30.8	8.8(4.8)	-	-
Dryopteris sp.	-	-	5.6	2.0	23.1	2.0(1.0)	-	-
Helodium blandowii	-	-	-	-	30.8	7.8(3.3)	-	-
Aulocomnium palustre	-	-	-	-	30.8	3.8(2.9)	-	-
Scutellaria galericulata	-	-	22.2	1.8(0.5)	38.5	2.2(0.8)	-	-
Iris versicolor	-	-	-	-	30.8	5.3(0.5)	-	-
Myrica gale	-	-	11.1	16.5(19.8)	15.4	16.0(19.8)	100.0	92.5(3.5)
Spiraea alba	6.7	3.0	5.6	2.0	23.1	7.7(10.9)	100.0	8.5(9.2)
Eleocharis sp.	20.0	12.3(11.7)	11.1	15.0(0.0)	7.6	1.0	-	-
Eriophorum tenellum	-	-	27.8	2.2(0.8)	7.7	3.0	-	-
Carex limosa	-	-	27.8	12.4(8.4)	-	-	-	-
Abiotic Variables								
Mat Thickness (cm)		12.2(4.2)		43.7(6.5)		66.7(5.8)		57.5(3.5)
Sphagnum Thickness (cm)		0.0(0.0)		0.8(1.4)		13.3(4.7)		0.0
Depth to Water Table (cm)		0.0		2.9(2.4)		16.1(4.5)		6.0(1.4)
Water Depth Above Mat (cm)		12.7(10.0)		5.3(10.6)		0.0		0.0
Water Depth Below Mat(cm)		16.1(20.7)		23.4(13.0)		17.0(6.3)		27.0(2.8)
Total Depth(water + mat)(cm)		41.0(14.9)		79.4(6.7)		89.5(2.0)		89.5(2.1)
pH of Surface Water		6.3(1.2)		5.2(0.5)		4.5(0.5)		3.9(0.4)
Number of Vascular Plants		4.9(2.2)		10.7(3.3)		12.6(1.4)		11.5(2.1)
Typha Density(#/m^2)		6.9(11.0)		14.1(12.6)		16.6(6.3)		1.5(2.1)

sampling five 1 m^2 quadrats in each 30 x 30m plot. In addition, shoot density, plant height and base diameter of shoots were recorded for Typha.

RESULTS AND DISCUSSION

Based on the phytosociological analysis of cover data, the 48 quadrats were divided into four groups (Table 1):

Group I: Vegetation on quadrats 1-15 that were dominated by Lemna minor and Sparganium. Riccia fluitans, Typha, Utricularia intermedia, and Carex pseudocyperus were also present. Quadrats in this group are characterized by a thin mat (12.2 ± 1.1 cm), high pH (6.3 ± 0.3), and an average of 13 cm standing water above the mat.

Group II: Quadrats 16-33 had more plant

Table 2. Pearson's product moment correlation coefficient of the biotic and abiotic factors of the marsh.

	Mat thickness	Depth of Water table	pH	Total depth	Depth below mat	Temperature below mat	Typha height	Typha shoot density	Typha cover	Sparganium cover	Lemna cover	Sphagnum cover	No. vascular plants	No. bryophytes
Mat thickness	1.000													
Depth of water table	0.802	1.000												
pH	-0.633	-0.542	1.000											
Total depth of water	0.908	0.687	-0.428	1.000										
Depth below mat	0.048	-0.030	0.225	0.363	1.000									
Temperature below mat	-0.881	-0.682	0.680	-0.793	-0.059	1.000								
Typha height	0.367	0.189	-0.295	0.261	-0.362	-0.356	1.000							
Typha shoot density	0.360	0.167	-0.233	0.255	-0.306	-0.299	0.906	1.000						
Typha cover	0.280	0.110	-0.221	0.250	0.250	-0.224	0.831	0.939	1.000					
Sparganium cover	0.597	-0.372	0.503	-0.649	-0.226	0.554	-0.463	-0.494	-0.456	1.000				
Lemna cover	0.680	-0.483	0.418	-0.723	-0.328	0.673	-0.164	-0.071	-0.025	0.473	1.000			
Sphagnum thickness	0.749	0.965	-0.482	0.513	-0.069	-0.599	0.212	0.208	0.164	-0.310	-0.395	1.000		
No. of vascular plants	0.736	0.593	0.616	0.708	0.295	-0.660	0.133	0.103	0.076	-0.541	-0.692	0.507	1.000	
No. of bryophytes	0.727	0.745	-0.484	0.683	0.143	-0.752	0.123	0.066	-0.005	-0.384	-0.488	0.681	0.621	1.000

species than those in Group 1. They were more acidic (pH 5.2 ± 0.1), had thicker mats (43.6 ± 1.6 cm) and fewer quadrats had standing water above the mat surface. Five quadrats had Sphagnum cover. Frequency and dominance of Sparganium and Lemna was less, while there was an increase in cover of Typha, Carex canescens, C. limosa, Galium palustre, Potentilla palustris, Agrostis hymenalis, and other herbaceous species. Carex limosa is an indicator species in this group although it occurred in only five quadrats. Field observations indicated that Typha growth was stunted when there was a continuous cover of Sphagnum and/or Carex spp.

Group III: This group is characterized by the total absence of Sparganium and Lemna, and high cover percentages for Sphagnum, Calamagrostis and many herbaceous plants. Typha, although present in every quadrat, was stunted. There was no standing water above the mat, the mat thickness averaged 66.7 ± 1.6 cm, and the group had low surface water pH (4.6 ± 0.1). Fen and bog forming species like Sphagnum squarrosum, Carex limosa, and Drosera rotundifolia were also present in the quadrats.

Group IV: Mat thicknesses in those two quadrats was nearly similar (57.5 ± 2.5 cm) to those in Group III, but the surface water was more acidic (pH 3.8 ± 0.2). The dominant species was Myrica gale and Typha glauca was absent. Except for Myrica, the species composition was more or less similar to Groups II and III (Table 1).

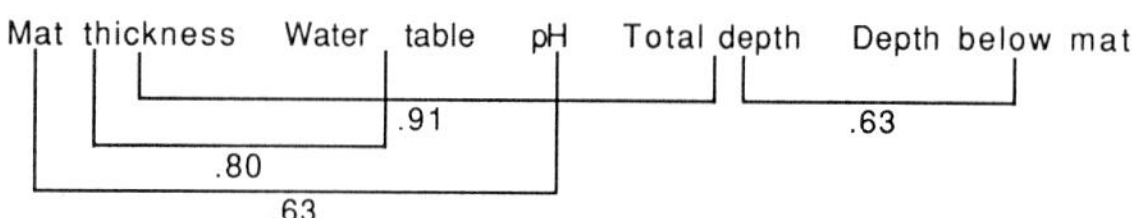

Figure 2. Pairwise Pearson product moment correlation among the independent variables (environmental factors).

Results of the analysis using Pearson's Product Moment Correlation Coefficient of the biotic and abiotic factors are presented in Table 2. Mat thickness is positively correlated with depth to water table, total depth of water, Sphagnum thickness, number of vascular plants and number of bryophytes. Water table depth was negatively correlated with pH, water depth below mat, Sparganium cover and Lemna cover.

Depth to water table on the mat is positively correlated with Sphagnum cover and number of bryophytes. Typha cover, shoot density and height were poorly correlated with mat thickness and pH.

Pairwise correlations among the environmental factors indicate that there is a very high correlation between the mat thickness and total depth (Fig. 2). The second and third best correlations were between mat thickness and water table, and mat thickness and pH, respectively.

Table 3. Strongest correlations between the dependent variables with two predictor (independent) variables.

Dependent Variables (plant factors)	Two Predictor Variables (environmental factors) with Highest Correlation			
	Mat thickness	Water table	Total depth	Depth below water
Typha height	.367			-.362
Typha shoot density	.360			-.306
Typha cover	.280		.250	
Sparganium cover	.597		-.649	
Lemna cover	.680		-.723	
Sphagnum cover	.749	.965		
No. vascular plants	.736		.708	
No. bryophytes	.727	.745		

Table 3 presents values of two independent variables that correlate most strongly with a particular dependent variable. It shows that the mat thickness is the single most important predicator parameter. The same result is shown in Table 4 in which environmental factors are correlated with the dependent variables (plant factors). The number of vascular plants was positively correlated with three environmental variables indicating that changes in mat thickness cause changes in the composition of vascular plants.

Relationships between pH, mat thickness, Sphagnum thickness and depth to water table of the quadrats are summarized in Fig. 3a-c. There is a clear positive correlation between the thickness of Sphagnum layer and depth to water table and a negative correlation with pH and mat thickness. There is a high correlation between mat thickness and total depth. The second and third best correlations were between mat thickness and water table, and mat thickness and pH, respectively.

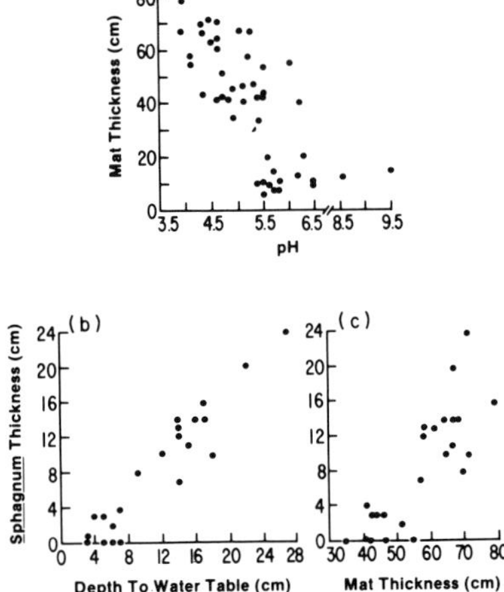

Figure 3. Relationships between (a) mat thickness and pH; (b) Sphagnum thickness and depth to water table; (c) Sphagnum thickness and mat thickness (from Mallik 1989).

Fig. 4 shows that the first Canonical of the plant factors are strongly related to the first Canonical variables of the environmental factors. The Canonical correlation was .98. When the first Canonical variable of the plant factors was plotted in relation to mat thickness, mat thickness was strongy correlated with plant factors (species structure and composition) (Fig. 5). All environmental variables together strongly correlated with the number of vascular plants and bryophytes, Sphagnum, Lemna and

Table 4. Strongest correlations between the independent variables (environmental factors) with dependent variables (plant factors).

Predictor (independent) variables (environmental factors)	Dependent Variable (plant factors) Correlate Most Strongly with Predictor Variable	
Mat thickness	Sphagnum cover .749	No. vascular plants .736
Water table	Sphagnum cover .965	No. Bryophytes .745
pH	No. vascular plants .616	Sparangium cover .745
Total depth	Lemna cover -.723	No. vascular plants .708
Depth below mat	Typha height -.362	Lemna cover -.328

Sparganium cover but correlate weakly with Typha parameters (height, shoot density, cover).

Species Distribution in Relation to Mat Thickness

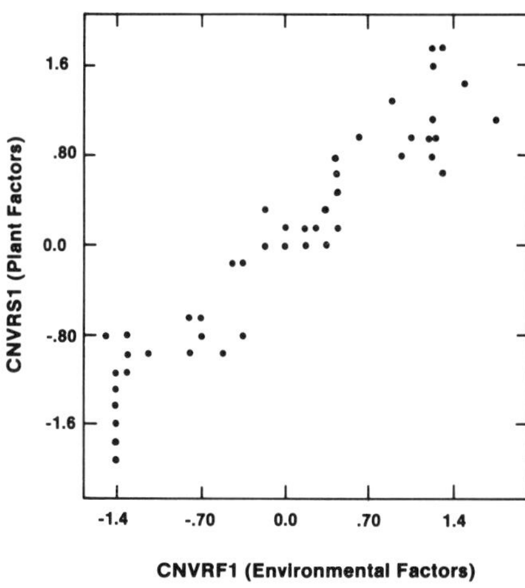

Figure 4. Relationship between the first canonical of the environmental factors with that of plant factors.

Figs. 6a-e show the distribution of cover value of several plants in the study area plotted against mat thickness. Several species (i.e. Sparganium (Fig. 6a) Lemna minor, Utricularia intermedia, Cicuta bulbiflora) are restricted to mats between 6 and 20 cm thick and with

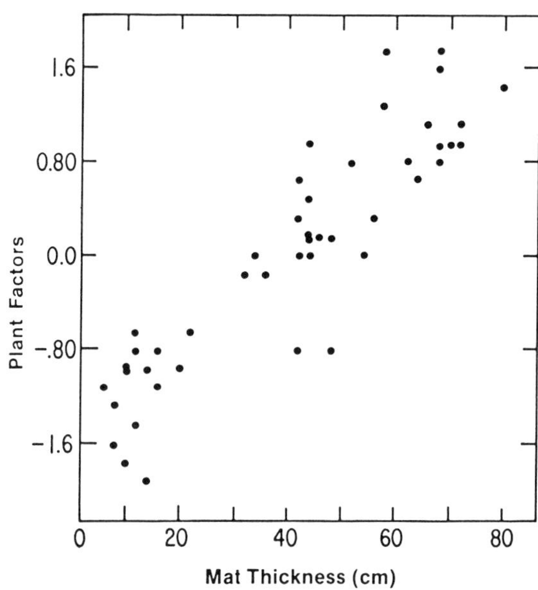

Figure 5. Relationships between mat thickness and plant factors of first canonical variable (from Mallik 1989).

standing water. Most of the other species were found on thicker mats (40-80 cm) with species like Calamagrostis canadensis, Viola macloskeyi, Menyanthes trifoliata, Pseudobryum cinclidioides, Lophocolia heterophylla, occurring on very thick mats (55-80 cm). Typha distribution was not related to mat thickness (Fig. 6b) although field observations indicate that it grows better on thinner mats or

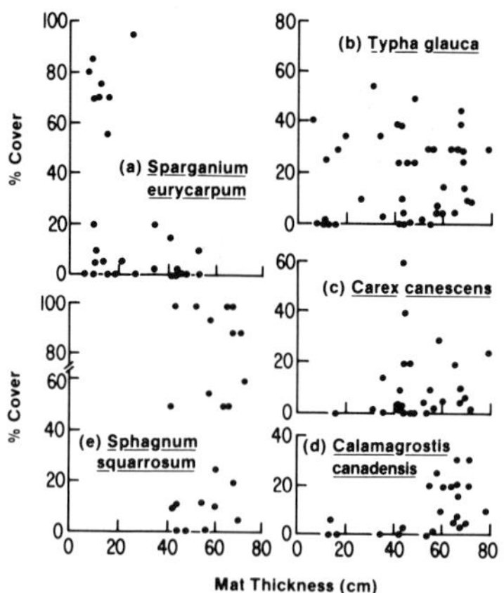

Figure 6. Relationship between plant cover and mat thickness for wetland species (from Mallik 1989).

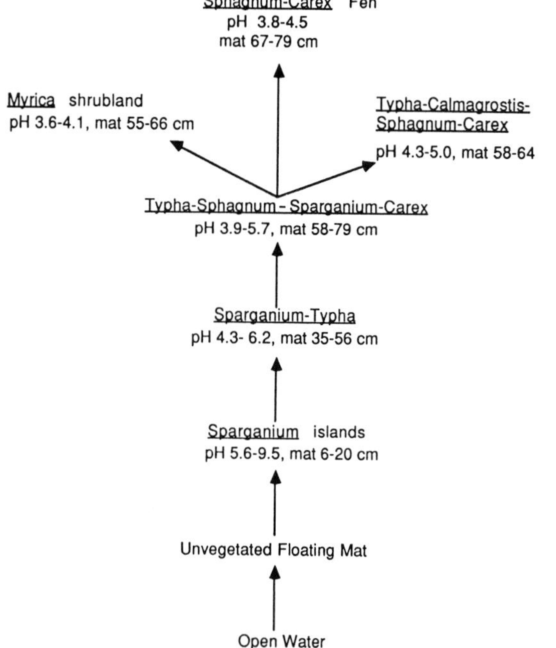

Figure 7. Postulated successional events in Tintamarre marsh.

on mats with standing water.

Examination of mat cores in the field as well as in the laboratory indicates that more than 90% of the material is composed of the roots, rhizomes, and leaves of Typha. In many cores examined in the field, seeds of Sparganium eurycarpum were found in the bottom 5-10 cm of the mat. The cores in the Sphagnum dominated area had 6-8 cm of live Sphagnum underneath. However, no trace of Sphagnum was found below that depth suggesting that its invasion is quite recent.

As stated in the Site Description, the southern part of the marsh contains open water with some isolated floating mats. Some of these mats had surfaced recently and thus were devoid of any vegetation. Two or three years after floating, the mats become dominated by Sparganium. At this stage pH is relatively high (6.5-8.6). As the mats grow older Typha invasion takes place and the habitat becomes more acidic. Typha dominates and biomass begins to accumulate due to the slow rate of decomposition. Typha dominance persists until the increased mat thickness and buoyancy leads to acidic, more nutrient-poor conditions. At this stage Typha vigour declines and bog forming species such as Sphagnum squarrosum, S. capillifolium, Carex limosa, Drosera rotundifolia, etc. appear, and in time dominate the habitat. Succession then may proceed in three directions: a) Typha may remain in the habitat in reduced vigour along with Sphagnum, Calamagrostis and a few Carex and herbaceous species; b) Sphagnum will dominate with some fen/bog species of Carex and Drosera; c) Myrica and Spiraea may invade, turning the habitat into a shrubland (Fig. 7).

It is clear that plant succession is closely linked with increases in mat thickness and buoyancy. Figure 3b, shows that with increase in mat depth, the thickness of Sphagnum layer increases along with the depth to water table from the mat surface. In the origin of bogs, the role of ground water hydrology is of considerable importance in the European mid-continent (Kulczynski 1949), in eastern Europe (Boelter and Verry 1977) and in the large peatland complexes of northwestern Minnesota and eastern North America (Glaser et al. 1981; Glaser and Janssens 1986; Janssens and Glaser 1986).

The postulated sequence of succession at Tintamarre marsh (Fig. 7) is also supported by the evidence from mat composition. In Sphagnum dominated areas, Sphagnum is underlain by a thick layer of Typha peat (mostly roots and rhizomes) even in areas where living Typha is absent. Occasional presence of the seeds of Sparganium eurycarpum in the lowermost layer of Typha mat seems to suggest that this species was abundant in the early part of succession. However, more quantitative data on mat composition would give further credence to the postulated successional sequence. This is likely to be difficult and time consuming, as structures are often similar between species (e.g., Typha versus Sparganium) and material deep in the mat is partially decomposed.

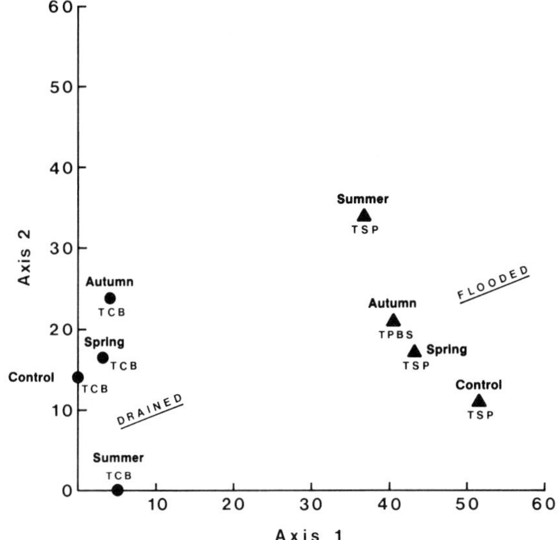

Figure 8. Strand ordination of the species cover data following drainage and seasonal burning. Species are arranged according to the decreasing cover value, e.g., TSP indicates the Typha had the highest cover, followed by Sphagnum and Pellia. T, Typha; C, Calamagrostis; B, Brachythecium; S, Sphagnum; P, Pellia (from Mallik and Wein 1986, Courtesy of NRC and Canadian Journal of Botany).

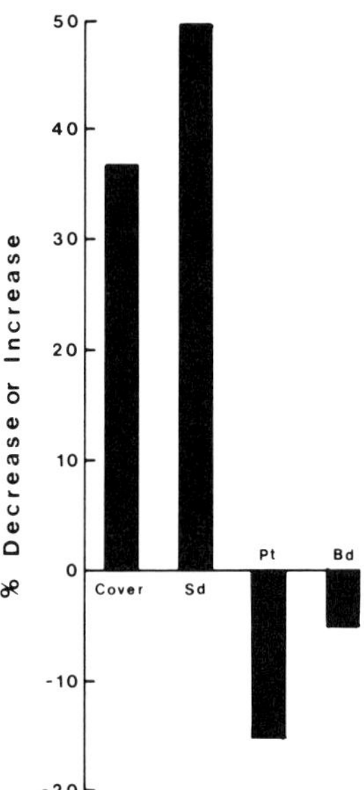

Figure 9. Percent decrease or increase of Typha cover, stem density (Sd), plant height (Pt), and base diameter (Bd) following draining and flooding treatments (results given as control on drained marsh as percentage of control on flooded marsh (from Mallik and Wein 1986).

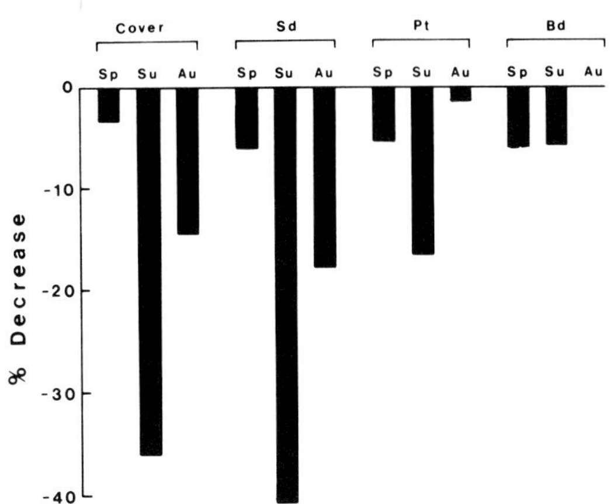

Figure 10. Percent decrease of Typha cover, stem density (Sd), plant height (Pt), and base diameter (Bd) following draining and seasonal burning treatments. Sp. spring; Su, summer; and Au, autumn burns (Mallik and Wein 1986 courtesy of NRC and Canadian Journal of Botany).

The patchiness of the fen and bog complex suggest that microscale differences have resulted in local differences in the rate and perhaps also in the direction of succession. Studies on microsite characteristics and associated vegetational changes in the initial stages of fen formation would provide a better understanding of the mechanism of development and functioning of the system.

Response of Overall Plant Community to Draining and Seasonal Burning

Draining has a significant effect on community composition as the total number of species per m^2 increased after three years. Cover and frequency of Aster novi-belgii, Lycopus uniflorus, Epilobium watsonii, Brachythecium salebrosum, Pleurozium schreberi, and Cladonia cristatella increased appreciably on the drained side whereas Carex spp., Lysimachia terrestris, Epilobium palustre, Pellia epiphylla, Sphagnum squarrosum, Drepanocladus exannulatus, and Helodium blandowii increased on the flooded side (for more detail see Mallik and Wein (1986). An ordination of cover value of all species clearly separated drained from flooded sites (Fig. 8) indicating that draining had a stronger control than burning. Except for Typha, which was always present in all quadrats, there were species differences between drained and flooded sites. In drained plots, Calamagrostis canadensis and Brachythecium salebrosum had high coverage whereas on the flooded side Sphagnum squarrosum and Pellia epiphylla were abundant and had high coverage.

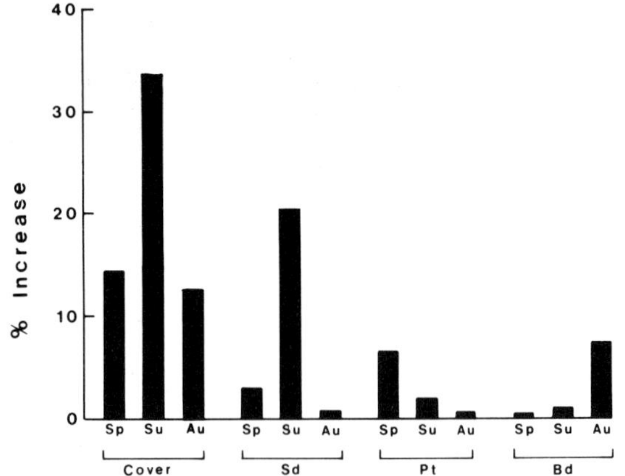

Figure 11. Percent increase of cover, stem density (Sd), plant height (Pt), and base diameter (Bd) of Typha following seasonal burning on the flooded side. Sp, spring; Su, summer; and Au, autumn burns (Mallik and Wein 1986 courtesy of NRC and Canadian Journal of Botany).

Typha maintained the highest frequency and cover in all plots and Viola macloskeyi, Potentilla palustris, Triadenum virginicum changed very little due to treatments. On the drained plots however, grass seedlings were more abundant compared to that in plots of the flooded side.

Response of Typha

Draining caused a significant increase in Typha coverage and shoot density, but a decrease in plant height and base diameter (Fig. 9). Response of Typha to seasonal burning on the drained and flooded sites are summarized in Fig. 10 and 11. On the drained sites, there was a decrease in all four parameters (Fig. 10). In contrast, seasonal burning on the flooded site caused an increase in all the above parameters (Fig. 11). The largest decrease in cover was obtained due to summer burn. On the drained site decreases in plant cover were 36, 14, and 3% due to summer, autumn, and spring burns, respectively.

Stem density on the drained site was decreased by 41, 18, and 7% due to summer, autumn, and spring burns, respectively. On the flooded site there were increases in shoot density by 20, 3, and 1%. Plant heights on the drained sites were decreased as a result of seasonal burning, while they increased on the flooded sites. Basal shoot diameter also decreased slightly due to spring and summer burns, while flooding and burning caused slight increases. However, the increases were not significant at the 5% level. Summer burning had the greatest impact on all variables except shoot base diameter on the drained site. The pattern is not as clear in the flooded site.

Management Implications

Gorham et al. (1979) reported that a rapid accumulation of organic matter from Typha (roots, rhizomes, stems, leaves, etc.) reduces the water depth and thus changes the physical and chemical characteristics of the marsh which favours paludification processes. Evolution of CO_2 and methane from anaerobic decomposition of organic matter makes the mat more buoyant and acidic, increasing the rate of decomposition and mineralization. From a management point of view, reduction in Typha cover is necessary in order to create conditions where open water is dispersed with patches of emergent species (Bishop et al., 1979; Murkin et al. 1982). Typha is an emergent species but it typically occurs in almost monospecific stands (Grace and Wetzel 1981). In spring and early summer, nutrients would be mobilized from the rhizome system to the aerial parts of the plant; thus, periodic draining plus summer burning and reflooding would be most effective in controlling Typha. Subsequent reflooding would increase species diversity by allowing germination and/or vegetative reproduction of both drawdown species and standing-water species (van der Valk 1981); many of which are related to the food chain of the wildlife in the marsh.

ACKNOWLEDGEMENTS

I thank Dr. R.W. Wein for useful discussions during the course of the work, Mr. H.R. Hinds and Dr. R.R. Ireland identified some angiosperms and bryophytes, respectively. Sincere help of Dr. E.H. Hogg and C. Hogg during field work is gratefully acknowledged. Dr. R. Bartlett and M.A. El-Bayomi helped in statistical analysis and computing, Dr. P. A. Thomas, B.A. Roberts, E.D. Wells, and Dr. E.H. Hogg made useful comments on the manuscript. The research was conducted during the tenure of a post-doctoral fellowship at the Fire Science Centre and Department of Biology, University of New Brunswick, Fredericton, N.B., Canada.

REFERENCES

Bishop, R.A., Andrew, R.D. and Bridges, R.J. 1979. Marsh management and its relation to vegetation, waterfowl and muskrats. Proceedings of the Iowa Academy Science 86: 50-56.

Boelter, D.H. and Verry, E.S. 1975. Peatland

and water in northern lake states - U.S.D.A. Forest Service, General Technical Report No. NC-31, 22 p.

Braun-Blanquet, J. 1932. Plant Sociology - the study of plant communities (translated, revised and edited by G.D. Fuller and H.S. Conard, 1972), Hafner Publishing Co., New York, NY. USA.

Cooley, W.W. and Lohnes, P.R. 1971. Multivariate Data Analysis. Wiley, New York, NY. USA.

Ganong, W.F. 1903. The vegetation of the Bay of Fundy salt and dyked marshes. An ecological study. Botanical Gazette 26: 161-186, 280-302, 349-367, 429-455.

Gauvin, J.M. 1979. Étude de la vegetation des marais sales du Parc National de Kouchibougauc, N.B., M.Sc. Thesis, Université de Moncton, Moncton, New Brunswick, Canada. 248 p.

Glaser, P.H. and Janssens, J.A. 1986. Raised bogs in North America: transitions in landforms and gross stratigraphy. Canadian Journal Botany 64:395-415.

Glaser, P.H., Wheeler, G.A., Gorham and Wright, H.W. 1981. The peat land mires of the Red Lake Peatland, Northern Minnesota: Vegetation, water chemistry, and landforms. Journal of Ecology 69: 575-599.

Gorham, E., Vitousek, P.M. and Reiners, W.A. 1979. Regulation of chemical budgets over the course of terrestrial ecosystem succession. Annual Review Ecology and Systematics 10: 53-84.

Grace, J.B. and Wetzel, R.G. 1981. Habitat partitioning and competitive displacement in cattails (Typha): experimental field studies. American Naturalist 118: 463-474.

Hale, M.E. 1961. Lichens Handbook. A guide to the Lichens of Eastern North America. Smithsonian Institute, Publication No. 4434, Washington, DC. USA. 178 p.

Hogg, E.H. and Wein, R.W. 1987. Growth dynamics of floating Typha mats: seasonal translocation and internal deposition of organic material. Oikos 50: 197-205.

Ireland, R.R. 1982. Moss Flora of the Maritime Provinces. National Museum of Natural Sciences, National Museums of Canada, Publications in Botany, No. 13. Ottawa, Ontario, Canada. 738 p.

Janssens, J.A. and Glasser, P.H. 1986. The bryophyte flora and major peat-forming mosses at Red Lake peatland, Minnesota. Canadian Journal of Botany 64: 427-442.

Kulczynski, S. 1949. Peat bogs of Polesie - Memoirs Academy Polish Sciences Serial B. No. 15, 356 p.

Mallik, A.U. and Wein, R.W. 1986. Response of a Typha marsh community to draining flooding and seasonal burning. Canadian Journal of Botany 64: 2136-2143.

Mallik, A.U. 1989. Small-scale plant succession towards fen on floating mat of a Typha marsh in Atlantic Canada. Canadian Journal of Botany 67: 1309-1316.

Murkin, H.R., Kaminski, R.M. and Titman, R.D. 1982. Responses by dabbling ducks and aquatic invertebrates to an experimentally manipulated cattail marsh. Canadian Journal of Zoology 60:2324-2332.

Roberts, B.A. and Robertson, A. 1986. Salt marshes of Atlantic Canada: their ecology and distribution. Canadian Journal of Botany 64: 455-467.

Rowe, J.S. and Scotter, G.W. 1973. Fire in the boreal forest. Quaternary Research 3: 444-464.

Scoggan, H.J. 1978-1979. Flora of Canada. Rt 1-4, National Museum of Natural Science, Ottawa, Canada.

Simard, A.J. 1975. Wildland fire occurrence in Canada. Canada Department of the Environment, Canadian Forestry Service, Ottawa, Canada.

Teal, J. and Teal, M. 1969. Life and death of the salt marsh. Little, Brown & Co., Boston. MA, USA. 278 p.

van der Valk, A.G. 1981. Succession in wetlands: A Gleasonian approach. Ecology 62: 688-696.

Wein, R.W. and Krusi, B.O. (eds.). 1982. A report on the second year of the Tintamarre marsh fire ecology study. pp. 46-51. Fire Science Centre, University of New Brunswick, Fredericton, N.B., Canada.

Wells, E.D. 1981. Peatlands of eastern Newfoundland: distribution, morphology, and nutrient status. Canadian Journal of Botany 59: 1978-1997.

THE RESPONSE OF DISTICHLIS SPICATA (L.) GREENE AND SPARTINA PATENS (AIT.) MUHL. TO NITROGEN FERTILIZATION IN HYDROLOGICALLY ALTERED WETLANDS

Dennis F. Whigham
Smithsonian Environmental Research Center
Edgewater, MD 21037

and

Sarah M. Nusser
Department of Statistics
Iowa State University
Ames, IA 50011

ABSTRACT

A fertilization study was conducted to test the hypothesis that substrate moisture content affects nitrogen uptake by plants in brackish wetlands that have been ditched for mosquito control. Spartina patens (Ait.)Muhl. and Distichlis spicata (L.)Greene were the test species. Nitrate fertilizer was applied to plots in an unditched (Control) area and in two areas that had been extensively ditched. Tissue nitrogen concentrations increased significantly in fertilized plots and they were significantly higher in the area where the substrate moisture content had decreased following management. Distichlis had a greater relative response to fertilization but the response did not differ between sites. Spartina, the more shallow rooted species, had the greatest relative response to fertilization in the driest area. The results demonstrate that minimum changes in plant tissue nitrogen concentrations can be achieved in ditched wetlands by maintaining water tables as near to the wetland surface as possible.

INTRODUCTION

Open Marsh Water Management (Ferrigno and Jobbins 1968), hereafter referred to as OMWM, and modifications of the original technique (Meredith et al. 1985) are widely used to control mosquito populations in coastal wetlands of the USA (Balling et al. 1980; Roman et al. 1984). The technique involves ditching the wetlands to destroy the breeding habitat of the mosquitos. OMWM does not appear to have any long-term negative impacts on vegetation (Shisler and Jobbins 1977) or invertebrates (Clarke et al. 1984) in frequently flooded coastal wetlands dominated by Spartina alterniflora. There may be negative impacts on birds that feed in salt marsh pools in Spartina alterniflora dominated coastal wetlands but other types of birds are not affected (Clarke et al. 1984).

In contrast, OMWM and related types of management can have negative impacts on the vegetation of brackish wetlands which only flood during high spring tides or storm tides if hydrologic conditions are not carefully controlled (Whigham et al. 1982; Roman et al. 1984). The key to successful implementation of OMWM in infrequently flooded brackish wetlands is to lower the water table enough to drain mosquito breeding depressions, but not far enough to permit the wetland to be invaded by undesirable shrub species such as Iva frutescens L. and Baccharis halimifolia L. (Meredith et al. 1985).

Lowering the water table on infrequently flooded brackish wetlands can also alter the nitrogen status of the vegetation. Whigham et al. (1982) found that the nitrogen content of above ground vegetation was significantly higher in areas where the water table had been lowered compared to areas where the water table had been maintained near the wetland surface.

In this paper we report results of an experiment to test the hypothesis that the increased shoot nitrogen concentrations measured by Whigham et al. (1982) was positively related to a decrease in the substrate moisture content. The experiment was designed in accordance with two postulates. First, that the addition of nitrogen fertilizer to any area affected by OMWM would result in an increase in shoot nitrogen concentrations. This assumption is based on the fact that plants in tidal wetlands respond positively to nitrogen fertilization (Broome et al. 1983; Chalmers 1979; Gallagher 1975; Valiela et al. 1982). The second assumption was that the relative response to fertilization would be greatest in areas where the substrate moisture content had been reduced by lowering the groundwater

D. F. Whigham et al. (eds.), Wetland Ecology and Management: Case Studies, 31–38.

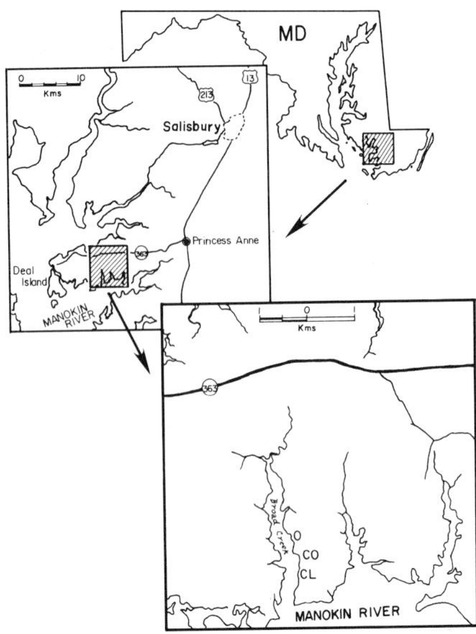

Figure 1. Location of Deal Island study sites. O= Open Site, CL= Closed Site, CO= Control Site.

table. The substrate in those areas should be more oxidized and more nitrogen should be available for assimilation by plants (Linthurst and Seneca 1981; Mendelssohn and Seneca 1980; Morris 1980).

METHODS

Study Location

The study was conducted on the Deal Island Wildlife Management Area on the Eastern Shore of Maryland, USA (Fig. 1). Wetlands within the management area were physically altered in 1979 as part of a study to determine the effects that different ditching and water management procedures had on controlling mosquitoes and on the structure and function of the wetlands (Lesser 1982; Whigham et al. 1982 and 1983).

This study was conducted in two of the areas that were ditched in 1979 and an 11.1 ha area, hereafter referred to as the Control Site, that had not been ditched. One of the ditched areas, hereafter referred to as the Open Site, was 20.2 ha in extent and was connected to the estuary by two ditches to permit tidal exchange. The hydrologic modifications at the Open Site resulted in a lowering of the water table and the greatest changes (5 to 10 cm) occurred within 10 meters of the ditches (Lesser 1982). The second ditched site, hereafter referred to as the Closed Site, was 12.1 ha in extent and was similar to the Open Site except that the ditch system was not connected to the estuary. Because there was no regular tidal exchange between the Closed Site and the estuary, the water table was not lowered. All three areas were dominated by Distichlis spicata (L.)Greene and Spartina patens (Ait.)Muhl. before ditching began. The species will hereafter be referred to as Spartina and Distichlis. Distichlis and/or Spartina dominated both areas after ditching but the Open Site was invaded by Iva frutescens L. and Baccharis halimifolia L. during the first growing season following the ditching (Whigham et al. 1983).

In June 1981, two sets of two plots, each 2 X 10 m, were randomly located at the Closed and Open Sites. One plot of each set was randomly located immediately adjacent to a ditch with the long axis parallel to the ditch edge. The second plot of the set was established with the same orientation but with the 2 meter dimension located between 18 and 20 m away from the side of the first plot. At the Control Site, two separate plots were used; the locations of the two plots were randomly determined. All ten plots were divided into 80 quadrats (0.5 m X 0.5 m) arranged in a 4 X 20 grid.

One set of plots was fertilized at the Open and Closed sites along with one of the plots at the Control Site. The plots were fertilized with NO_3-N at a rate of 20 gN m^2 on 11 June (1981), a time of rapid shoot growth. The 100% NO_3-N fertilizer (Sudbury Laboratory; Sudbury, Massachusetts) was dissolved in estuarine water prior to application. The loading rate was similar to that used by Gallagher (1975) and falls within the range of nitrogen loading to various estuarine wetlands (Valiela et al. 1985).

Shoots of Distichlis and Spartina were collected from three randomly selected quadrats in each plot after 1, 2, 3, 6, and 9 weeks following addition of the fertilizer. The shoots were dried at 60 °C, ground in a Wiley Mill to pass through a 2 mm screen, and analyzed for nitrogen using Kjeldahl procedures (APHA 1976).

Substrate moisture content was measured on triplicate cores that were collected in June (11, 19, 25), July (9, 22) and August (3). At the Open and Closed sites, samples were collected between 0-2 and 18-20 m from randomly located positions along the ditches. Triplicate random samples were also collected from the Control Site. The aluminum coring tube was 20 cm long and had a diameter of 5 cm. Moisture content, (wet weight-dry weight)/wet weight, was determined by weighing the cores after they were extruded from the sampling tubes and after they had been dried at 60 °C.

The vertical distribution of live belowground biomass (roots and rhizomes) of Distichlis and Spartina was determined by collecting cores in August from areas which had approximately 50% coverage of both species, the most common situation in all plots. Triplicate cores, 20 cm deep and 5 cm in diameter, were collected from each area. The cores were extruded and cut into 5 cm sections that were processed using the procedures described by Gallagher (1975). Dry weights of root/rhizome material were determined for each species.

Data Analysis

Percent nitrogen data were arsine and square root transformed and then analyzed in two stages: (1) the Closed and Open Sites were compared to detect differences in plant responses to fertilization, distance from the ditch, and temporal patterns, and (2) comparisons were made between the Control Site and the Closed and Open Sites to investigate differences in responses to fertilization and time.

The analyses were complicated by the lack of true replication (i.e., the quadrats within a single 2 m X 10 m plot are subsamples). To circumvent this problem, sites were treated as blocks. In the first phase of the analysis, the experiment was then analyzed as a split-plot design where the whole plots, distance X fertilizer combinations, were arranged in a randomized complete block design and time was regarded as the split-plot. Interactions involving site X time were tested with sampling error.

For the second part of the analyses, differences between the Control and the Closed and Open sites for percent nitrogen were investigated using Bonferonni t-tests. Pairwise comparisons were made between Open and Control sites and between Closed and Control sites for each fertilizer X distance X time combination.

RESULTS

Spartina patens

There were significant site ($P < .001$) and fertilization ($P < .001$) effects on the nitrogen concentrations of Spartina shoots when the Open and Closed Sites were compared. Mean shoot nitrogen concentrations at the Open Site (mean ± 1 standard error for the four sampling periods: 1.34 ± 0.03%) were significantly higher than at the Closed Site (1.03 ± 0.04%) and plants in fertilized quadrats had significantly

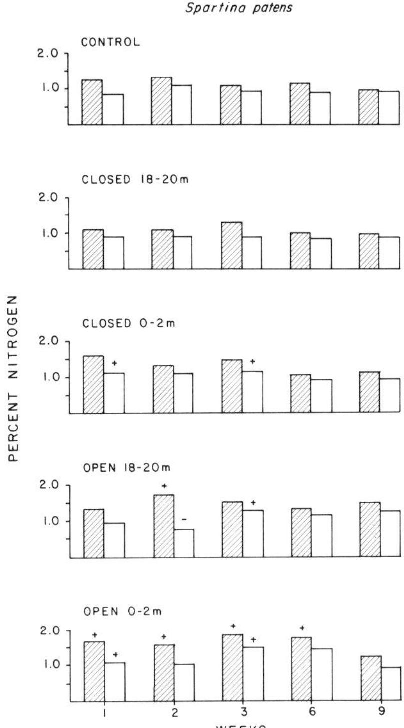

Figure 2. Mean shoot nitrogen concentrations (%) of S. patens in fertilized (shaded bars) and unfertilized (unshaded bars) quadrats at the Control, Open, and Closed Sites. Values are means. Means of values at the Open and Closed sites that are significantly different ($P < 0.05$) from means at the Control Site are indicated with either a + (Mean greater than plants from the Control Site) or - (Means less than plants from the Control Site.

higher %N (1.37 ± 0.04) than plants in quadrats that were not fertilized (1.04 ± 0.04). Percent nitrogen in shoots at 0-2 meters (1.23 ± 0.04) was not significantly different from percent N at 18-20 m (1.18 ± 0.04). There were no significant time effects or 2-way interactions. There were significant 3-way interactions among site X time X fertilizer ($P < 0.01$) and site X time X distance ($P < 0.01$), and the 4-way interaction was significant ($P < 0.01$). However, the main features of the data are probably the fertilizer and site effects (mean squares of .0063 and .0048, respectively) since using a conservative degrees of freedom test on all effects involving time (to account for correlation among observations on the same unit over time), showed that the significant time interactions were based on sampling error, which undoubtedly underestimated experimental error.

Figure 2 compares shoot nitrogen

concentrations at the Control Site with those at the two distances at the Open and Closed Sites. In all but one instance (Week 9), %N was significantly (P < .05) higher in fertilized quadrats near the ditch at the Open Site than the Control fertilized plots (Fig. 2). In contrast, plants from fertilized quadrats between 18 and 20 m at the Open Site had significantly higher shoot nitrogen concentrations relative to Control fertilized plots only on week 2. Relative to unfertilized Control plots, plants in unfertilized quadrats at the Open site had significantly higher nitrogen concentrations at 0-2 m on week 1 and 3 and at 18-20 m on week 3. There were few significant differences in shoot nitrogen concentrations when the Control and Closed Sites were compared. At the Closed Site, plants in unfertilized quadrats near the ditch had significantly (P < .05) higher %N than plants from the unfertilized plots at the Control Site at weeks 1 and 3.

Site differences in the magnitude of the fertilization response of Spartina ranged from 38.4 ± 10.7% at the Open Site to 26.4 ± 3.5% and 24.8 ± 7.5% at the Closed and Control sites respectively (Fig. 4)

Distichlis spicata

Site (P < .001), fertilization (P < .001), and distance (P < .05) effects were significant when %N in Distichlis shoots was compared at the Open and Closed sites. Nitrogen concentrations at the Open Site (1.49 ± 0.04%) were significantly higher than those measured at the Closed Site (1.34 ± 0.04%) and significantly higher near the ditches (1.45 ± 0.04%) than at 18-20 m (1.38 ± 0.04%). Shoots collected from fertilized quadrats had significantly higher %N (1.62 ± 0.04) than shoots from quadrats that were not fertilized (1.28 ± 0.04). There were several significant two and three way interactions involving time: time X fertilizer (P <.01); time X distance (P < .001); site X time (P < .01); site X time X fertilizer (P < .001); site X time X distance (P < .001). The four-way interaction was also significant (P < .001). Similar to Spartina, the main treatment effects are probably more important than the interactions involving time because the conservative degrees of freedom tests showed that the significant interactions with time were based on sampling rather than experimental error.

Comparisons between the Control Site and two ditched sites for D. spicata are shown in Fig. 3. Although there was a tendency for nitrogen concentrations to be elevated in fertilized plots at the Open and Closed Sites relative to fertilized plots at the Control Site,

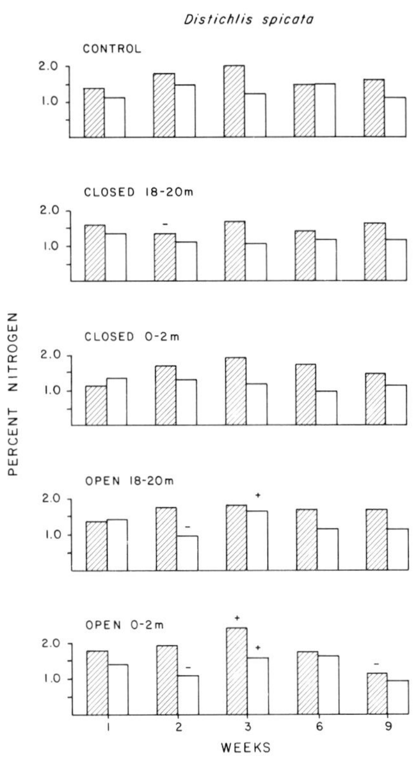

Figure 3. Mean shoot nitrogen concentrations (%) of D. spicata in fertilized (shaded bars) and unfertilized (unshaded bars) quadrats at the Control, Open, and Closed Sites. Values are means. Means of values at the Open and Closed sites that are significantly different (P < 0.05) from means at the Control Site are indicated with either a + (Mean greater than plants from the Control Site) or - (Means less than plants from the Control Site.

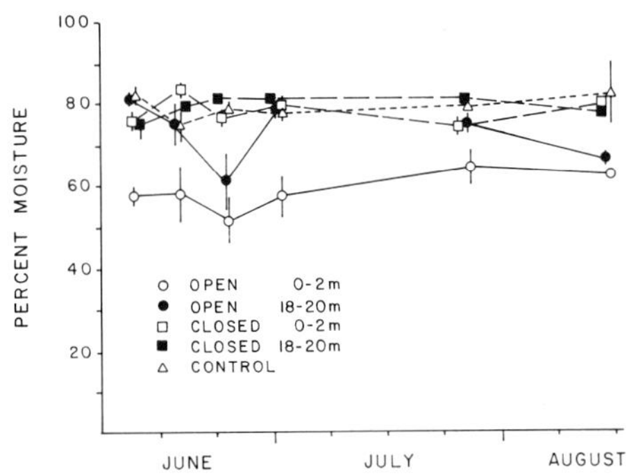

Figure 4. Percent change in the nitrogen concentration of S. patens and D. spicata shoots at the three study sites. Values are means (± 1 standard error) which were calculated as follows: ((%N in fertilized plots - %N in unfertilized plots) / %N in unfertilized plots) X 100. Site designations are as in Fig. 1.

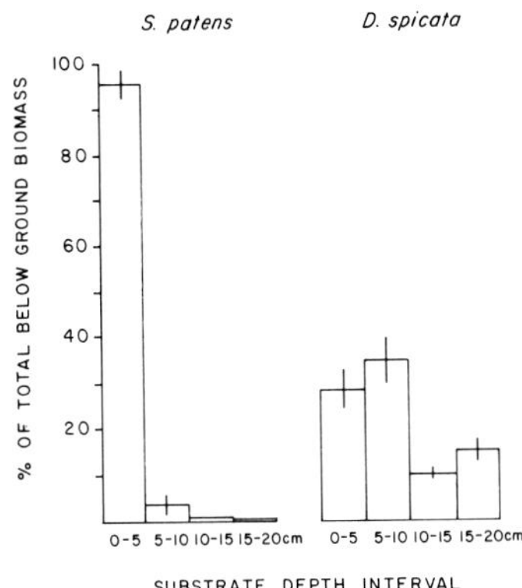

Figure 5. Substrate moisture content at the Open, Closed, and Control Sites. All values are means ± 1 standard error.

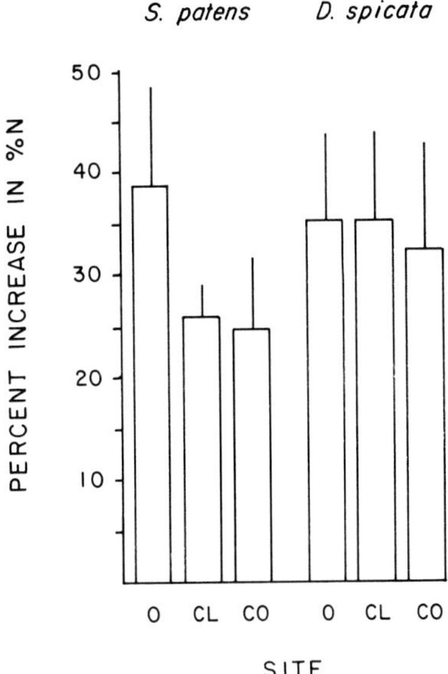

Figure 6. Distribution of roots and rhizomes of D. spicata and S. patens growing in areas where there is an approximately equal coverage of both species. All values for the 5 cm depth intervals are means of triplicate samples ± 1 standard error of the mean.

the values were only significantly higher ($P < .05$) than those measured at the Control Site on week 3 at 0-2 m. Shoots in fertilized plots at the Open and Closed sites had significantly lower %N in relation to the fertilized Control Site on two occasions (Closed Site: week 2 at 18-20 m and Open Site: week 9 at 0-2 m). Shoot %N in unfertilized plots at the Open Site on week 2 was significantly lower then the unfertilized plot at the Control Site.

As indicated, Distichlis responded positively to fertilization at all three sites (Fig. 4). The percent increase for Distichlis was, however, very similar at all three sites.

Substrate Moisture

Substrate moisture content was lowest near the ditch at the Open Site and there were no clear differences between the other sites and distances (Fig. 5). Moisture content ranged from 78.4 ± 0.10 to 83.1 ± 1.3% at the Control Site and from 75.4 ± 4.6 to 82.1 ± 1.1% at the Closed Site.

Distribution of Roots and Rhizomes

Most of the below ground biomass (95.2 ± 3.5%) of Spartina occurred within the first 5 cm of substrate (Fig. 6). The distribution of Distichlis roots and rhizomes was more uniform with only 27.9 ± 4.6% of the biomass in the top 5 cm of substrate.

DISCUSSION

The positive responses of Spartina patens and Distichlis spicata to fertilization are similar to those that have been demonstrated for a more frequently studied estuarine species, Spartina alterniflora (Sullivan and Daiber 1974; Broome et al. 1983; Gallagher 1975; Haines and Dunn 1976; Chalmers 1979; Smart and Barko 1980; Valiela et al. 1982 and 1985). The reasons for increased nitrogen utilization under fertilized conditions are complex (Mendelssohn et al. 1982) but most results indicate that this may be caused by an increase in the availability of ammonium. Our results indicate that a decrease in substrate moisture content can also results in an increase in the availability of ammonium.

Comparing sites, shoot nitrogen concentrations were significantly higher at the Open Site where the water table had been lowered (Lesser 1982) and where the substrate moisture content had decreased the most, especially near the ditches. That result supports the hypothesis that substrate waterlogging is one of the key factors that controls nitrogen uptake in infrequently flooded brackish wetlands (Mendelssohn et al. 1982). A decrease in substrate moisture content could coincide with an increase in substrate oxidation by several mechanisms. First, oxygen could pass from the atmosphere through plant aerenchyma and into the substrate by diffusing out of roots (Teal and Kanwisher 1966; Mendelssohn et al. 1982; Howes et al. 1981; Mendelssohn and Postek 1982). Plant evapotranspiration and tidal pumping are two other mechanisms by which oxygen can enter the substrate (Dacey and Howes 1984). Valiela et al. (1976) have also shown that diffusion of oxygen into the substrate can be augmented by fertilizer application which can cause increased evapotranspiration. Finally, increasing the rate of movement of interstitial water would also produce the same response as lowering the moisture content of the substrate. Wiegert et al. (1983) have shown that short-form S. alterniflora production increased when drainage tiles were placed in the substrate to increase the movement of subsurface water. The response measured by Wiegert et al. (1983) was, however, most likely caused by a more rapid removal of toxic sulfides and delivery of iron than any change in substrate moisture content (King et al. 1982) or alleviation of anaerobiosis in the rhizosphere. In our study, increased water movement within the substrate would have only occurred near the ditches at the Open Site and Lesser (1982) has shown that the groundwater table fluctuated most at those locations. The effects of evapotranspiration and diffusion would have been similar at both sites as there were not any differences in standing live biomass (Whigham et al. 1982 and 1983). Both factors would, however, have been augmented at the Open Site by the fact that the groundwater was rarely recharged by surface flooding events and the substrates - especially near ditches - drained during low tide. Drainage could have been primarily responsible for the significantly higher shoot nitrogen concentrations of Spartina in fertilized and unfertilized plots near the ditch at the Open Site.

There were also significant differences between the two species in their response to fertilizer addition. While Distichlis responded positively to fertilization, the percent increase in shoot nitrogen between fertilized and unfertilized plots was very similar at all three sites. The lack of a clear site effect for Distichlis may have been due to the fact that it roots deeper in the substrate where soil moisture levels were higher and similar at all sites. Ditching of the types of wetlands that we have studied would, therefore, have little effect on nitrogen assimilation by Distichlis unless the water table was lowered more. Site differences in the magnitude of the fertilization response of Spartina were greater at the Open Site, particularly near the ditch. This is consistent with the fact that Spartina is much more shallow rooted than Distichlis and lowering of the water table by as little as 5 cm could have a positive effect on nitrogen uptake by Spartina.

In this study we have demonstrated that wetland management for mosquito control can have an effect on nutrient cycling if the substrate moisture content is lowered, especially wetlands dominated by Spartina. We have shown elsewhere (Whigham et al. 1982 and 1983) that lowering the substrate moisture content can also produce dramatic changes in the vegetation. The same types of manipulations did not, however, have any negative impacts on water quality parameters in ditches and ponds nor on patterns of nutrient exchange between the managed wetlands and the estuary (Whigham et al. 1983). We agree with Meredith et al. (1985) who concluded that proper control of the water table will result in minimum disruption of ecological processes in infrequently flooded tidal wetlands that have been ditched.

ACKNOWLEDGEMENTS

We would like to thank Cavell Brownie (North Carolina State University) for statistical advice. Tom Jordan, Jay O'Neill, Carin Chitterling, Bert Drake, Irv Mendelssohn, Brian Palmer, Allison Snow, and Paul Honz provided editorial comments. The research was supported by the Smithsonian Environmental Sciences Program and the Maryland Department of Natural Resources, Tidewater Administration.

REFERENCES

American Public Health Association. 1976. Standard Methods for the Examination of Water and Wastewater, 14th edition. American Public Health Association, New York, NY. USA. 1193 p.

Balling, S.S., Stoehr, T. and Resh, V.H. 1980. The effects of mosquito control recirculation ditches on the fish community of a San Francisco Bay salt marsh. California Fish and Game 66: 25-34.

Broome S.W., Seneca, E.D. and Woodhouse, W.W. 1983. The effects of source, rate and placement of nitrogen and phosphorus fertilizers on growth of Spartina alterniflora transplants in North Carolina. Estuaries 6:212-226.

Chalmers, A.G. 1979. The effects of fertilization on nitrogen distribution in a Spartina alterniflora salt marsh. Estuarine and Coastal Marine Science 8: 327-337.

Clarke, J.A., Harrington, B.A., Hruby, T. and Wasserman, F.E. 1984. The effect of ditching for mosquito control on salt marsh use by birds in Rowley, Massachusetts. Journal of Field Ornithology 55: 160-180.

Dacey, J.W.H. and Howes, B.L. 1984. Water uptake by roots controls water table movement and sediment oxidation in short Spartina marsh. Science 224: 487-489.

Ferrigno, F. and Jobbins, S.M. 1968. Open marsh water management. Proceedings New Jersey Mosquito Extermination Association 55: 104- 115.

Gallagher, J. 1975. Effect of an ammonium nitrate pulse on the growth and elemental composition of natural stands of Spartina alterniflora and Juncus roemerianus. American Journal of Botany 62:644-648.

Haines, B.L. and Dunn, W.L. 1976. Growth and resource allocation responses of Spartina alterniflora Loisel. to three levels of NH_4-N, Fe and NaCl in solution culture. Botanical Gazette 137: 224-230.

Howes, B.L., Howarth, E.Q., Valiela, I. and Teal, J.M. 1981. Oxidation-reduction potentials in a salt marsh; spatial patterns and interactions with primary production. Limnology and Oceanography 26:350-360.

King, G.M., Klug, M.J., Wiegert, R.G. and Chalmers, A.G. 1982. Relation of soil water movement and sulfide concentration to Spartina alterniflora production in a Georgia salt marsh. Science 218: 61-63.

Lesser, C.R., 1982. A Study of the Effects of Three Mosquito Control Marsh Management Techniques on Selected Parameters of the Ecology of a Chesapeake Bay Tidewater Marsh in Maryland. Maryland Department of Natural Resources, Annapolis, MD, USA. 116 p.

Linthurst, R.A. and Seneca, E.D. 1981. Aeration, nitrogen and salinity as determinants of Spartina alterniflora Loisel growth response. Estuaries 4: 53-63.

Mendelssohn, I.A. and Seneca, E.D. 1980. The influence of soil drainage on the growth of salt marsh cordgrass Spartina alterniflora in North Carolina. Estuarine and Coastal Marine Science 11: 27-40.

Mendelssohn, I.A. and Postek, M.T. 1982. Elemental analysis of deposits on the roots of Spartina alterniflora Loisel. American Journal of Botany 69: 904-912.

Mendelssohn, I.A., McKee, K.L. and Postek, M.T. 1982. Sublethal stresses controlling Spartina alterniflora productivity. pp. 223-242. In Gopal, B., Turner, R.E., Wetzel, R.E. and Whigham, D.F., eds. Wetlands: Ecology and Management International Science Publications, Jaipur, India.

Meredith, W.H., Saveikis, D.E. and Stachecki, C.J. 1985. Guidelines for "Open Marsh Water Management" in Delaware's salt marshes objectives, system designs, and installation procedures. Wetlands 5: 119-133.

Morris, J.T. 1980. The nitrogen uptake kinetics of Spartina alterniflora in culture. Ecology 61: 1114-1121.

Roman,C.T. Niering, W.A. and Warren, R.S. 1984. Salt marsh vegetation change in response to tidal restriction. Environmental Management 8: 141-150.

Shisler, J.K. and Jobbins, D.M. 1977. Salt marsh productivity as effected by the selected ditching technique, Open Marsh Water Management Mosquito News 37: 631-636.

Smart, R.M. and Barko, J.M. 1980. Nitrogen nutrition and salinity tolerance of Distichlis spicata and Spartina alterniflora. Ecology 61: 630-638.

Sullivan, M.J. and Daiber, F.C. 1974. Responses in production of cordgrass, Spartina alterniflora, to inorganic nitrogen and phosphorus fertilizer. Chesapeake Science 15: 121-123.

Teal, J.M. and Kanwisher, J.W.1966. Gas transport in the marsh grass, Spartina alterniflora. Journal of Experimental Botany 17: 355-361.

Valiela, I., Teal, J.M. and Persson, N.Y. 1976 Production and dynamics of experimentally enriched salt marsh vegetation: Below-ground biomass. Limnology and Oceanography 21: 245-252.

Valiela, I., Howes, B.L., Howarth, R.W., Giblin, A.E., Forman, K., Teal, J.M. and Hobbie, J.M. 1982. Regulation of primary production and decomposition in a salt marsh ecosystem. pp. 245-252 In Gopal, B.,Turner, R.E., Wetzel, R.G. and Whigham, D.F., eds. Wetlands: Ecology and Management. International Science Publications, Jaipur, India.

Valiela, I., Teal, J.M., Cogswell, C., Hartman, J., Allen, A., Van Etten, R. and Goehringer, D. 1985. Some long-term consequences of sewage contamination in a salt marsh ecosystem. pp. 301-306. In Godfrey, P.J., Kaynor, E.R. and Pelczarski, S., eds. Ecological Considerations in Wetlands Treatment of Municipal Wastewater. Van Nostrand Reinhold Company, New York, NY. USA.

Wiegert, R.G., Chalmers, A.G. and Randerson, R.F. 1983. Productivity gradients in salt marshes: the response of Spartina alterniflora to experimentally manipulated soil water movement. Oikos 41: 1-6.

Whigham, D.F., O'Neill, J. and McWethy, M. 1982. Ecological implications of manipulating wetlands for purposes of mosquito control. pp. 459-476 In Gopal, B., Turner, R.E., Wetzel, R.G. and Whigham, D.F. Wetlands: Ecology and Management. International Science Publications, Jaipur, India.

Whigham, D.F., O'Neill, J. and McWethy, M. 1983. The Effect of Three Marsh Management Techniques on the Ecology of Irregularly Flooded Chesapeake Bay Wetlands. Maryland Department of Natural Resources, Annapolis, MD, USA. 295 p.

MAXIMIZING DUCKWEED (LEMNACEAE) PRODUCTION BY SUITABLE HARVEST STRATEGY

Eliska Rejmánková and Marcel Rejmánek
Department of Botany, University of California, Davis CA. 95616

Jan Kvet
Department of Hydrobotany, Institute of Botany,
Czechoslovak Academy of Science
379 82 Trebon, Czechoslovakia

ABSTRACT:

Duckweeds (Lemnaceae), rapidly growing floating aquatic plants with a high protein content, are becoming more and more promising in waste water treatment and/or as an animal feed. An attempt has been made to develop the optimal harvest strategy for obtaining the maximum yields from duckweed stands. Usually, only discrete harvests in several day intervals are feasible and the end solution of the problem is not entirely trivial. Under a given intrinsic growth rate determined by climatic conditions and non-limiting mineral resources, the yield depends on the initial biomass and on both the size and frequency of harvest. The multi-stage harvesting model of Elizarov and Svirezev (1972) based on the logistic growth function was used and the predictions were compared with the experimental data. The highest total yield of giant duckweed (Spirodela polyrhiza) obtained by arbitrarily chosen harvest regime fell short of the predicted maximum possible total yield, calculated using the same sequence of harvest intervals, by 27.2%.

INTRODUCTION

The role of duckweed in aquaculture, waste water management and production of animal feeds was thoroughly summarized by Culley et al. (1983). Several other papers pointed out the potential usefulness of duckweed in aquaculture (Porath and Pollock 1982, Cassani et al. 1982; Mbagwu and Adeniji 1988). Numerous projects have been carried out on investigating the possibilities of duckweed use on both experimental and pilot scale (DeBusk et al.1981: Reddy and DeBusk 1985a, 1985b).

Duckweed species are highly productive under favorable conditions (Ericson et al. 1982; Mestayer et al. 1984; Oron et al. 1985; Landolt 1987). Markedly density dependent development of their populations (Ikusima 1955; Clatworthy and Harper 1962; Rejmánková 1979) represents a textbook example of logistic growth.

Yet, very few experiments have been made, to our knowledge, to assess the optimum conditions for obtaining the highest yields by discrete harvesting. The optimum harvesting regime can be assessed in two ways: (a) empirically, as a choice of the best of many variants tested; (b) mathematically, if the growth function is known and estimates of its parameters can be made. In the majority of cases, the studies on duckweed harvesting were exclusively empirical. For example, DeBusk et al. (1981) assessed the optimum stocking density as 38 g m^{-2} and then designed their sampling so as to keep this density more or less constant throughout the whole experiment. Another approach used by Reddy and DeBusk (1985b) was to start with low density, (10g m^{-2}), allow plants to grow until maximum densities were reached and then harvest and restock to original low densities.

A typical example of results from a harvesting experiment is illustrated by our data in Fig. 1. The search for optimal harvest strategies in populations of other aquatic species has followed a similar trial-and-error pattern (Lorber et al. 1984; Madsen et al. 1988; Painter 1988; Reddy 1983; Tucker 1983).

Several attempts have been made to derive theoretically the discrete harvesting strategy for obtaining maximum total yield. Numerical approximations have mainly been used (e.g., Lorber et al. 1984; Seip 1980). The mathematical solution of the problem by Elizarov and Svirezev (1970, 1972) still appears to be little known in spite of its detailed review by Swan (1975). Their method can be used for calculation of the optimum harvest fraction as a function of the time interval between successive

D. F. Whigham et al. (eds.), Wetland Ecology and Management: Case Studies, 39–43.

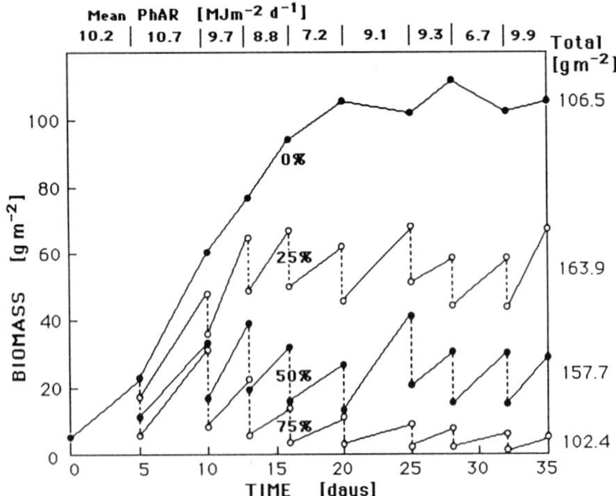

Figure 1. Experimental harvesting of 0, 25, 50, and 75% of giant duckweed (Spirodela polyrhiza) biomass (dry weight) at 3 to 5 day intervals for 35 days. The respective total yields are 106.5, 163.9, 157.7, and 102.4 g m^{-2}. The recorded interval means of daily PhAR are indicated at the top of the figure.

harvests, the biomass at the start of each interval, and expected intrinsic growth rate. Elizarov and Svirezev's method was originally formulated for continuous logistic growth but it can be modified for more general functions (Swan 1975).

THE ELIZAROV AND SVIREZEV APPROACH

Assume that population growth follows the logistic function

$$dW/dt = r W (K-W) /K \qquad (1)$$

where W denotes biomass at time t, r is the intrinsic growth rate of the biomass and K is the carrying capacity of the environment, i.e., a stable equilibrium of (1). With $W = W_0$, the initial biomass at time t_0, logistic equation (1) has the solution

$$W_t = K/[1 + (K/W_0 - 1)e^{-rt}] \qquad (2)$$

If growth of a population follows (1) and r is not constant because the environment may be, in general, different in each interval between successive harvests, the maximum yield is attainable if the harvest fractions, k_i ($0 < k_i < 1$) satisfy

$$k_i = 1 - K/[W_{0i}(e^{rh/2} + 1)] \qquad (3)$$

where h is the time interval between successive harvests (i, i+1) and W_{0i} is the biomass at the beginning of an i-th interval. The relation of k to h, for selected r values, is shown graphically in Fig. 2. It is important to realize that if $k_i < 0$, then W_{0i} or r or both are too small and biomass should not be harvested at all or h should be increased. In our simulation, optimal harvest fractions k_i were calculated (a) on the assumption that the photosynthetically active radiation (PhAR), and consequently r, will be the same in the subsequent between-harvest interval as they were in the preceding interval, and (b) on the basis of actual PhAR in the subsequent-harvest interval.

If r is approximately constant during the time horizon, T, the maximum total yield, Y_{max} obtainable through multiple-stage harvesting is

$$Y_{max} = rW_{01}\lambda / [r+ (r/K) W_{01} (\lambda -1)] + [(T-h/h] K (\lambda^{1/2} - 1) / (\lambda^{1/2} + 1), \qquad (4)$$

Where $\lambda = e^{rh}$ and W_{01} is biomass at the beginning of the first interval (see Fig. 3). It can be shown that as H $\rightarrow$ 0, Ymax $\rightarrow$ $W_{01}+r^2KT/4r$. This is the maximum yield under continuous harvesting - a strategy which is, of course, technically not feasible for macrophytes. The yields achieved in reality can be compared with corresponding values of Y_{max}. Figure 3 clearly shows that for small intrinsic growth rates (r) the length of time intervals between harvests is not so important; for higher r values, choice between intervals of say four or eight days could make a substantial difference. For detailed treatment of (3) and (4) see Elizarov and Svirezev (1970, 1972) and Swan (1975).

EXPERIMENTS AND SIMULATION

The experiments were carried out with the giant duckweed, Spirodela polyrhiza (L.), Schleid., in a small fishpond (about 2 ha) at Mokriny by Trebon, South Bohemia, Czechoslovakia (49°N, 14°45'E, altitude 430 m, climate is temperate, sub-oceanic). The fishpond receives waste water from a duck hatchery. The average inorganic nitrogen (mostly in the form of NH_4) and phosphorus (PO_4) content in its water was 50 and 8 mg l^{-1}, respectively. The growth of duckweed is known to be independent of concentration of N and P higher than 10 and 2 to 3 mg l^{-1}, respectively (Rejmankova 1982). The duckweeds were grown within floating wooden

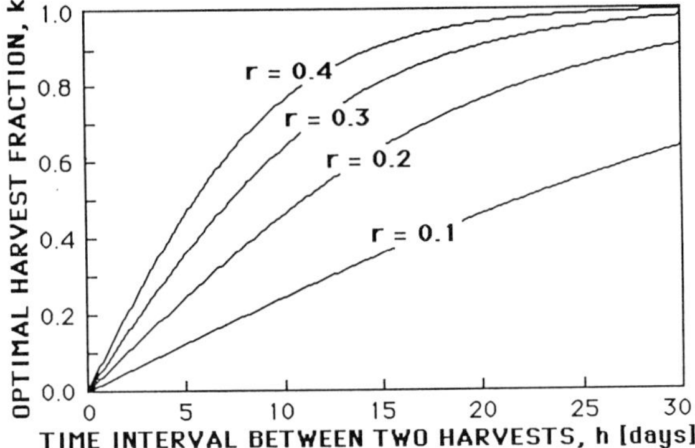

Figure 2. The optimal harvest fraction, k, as a function of the time interval between two harvests (h) for four different specific growth rates, r (g g^{-1} d^{-1}), assuming logistic growth, W_0 = 50 g m^{-2}, and K = 100 g m^{-2}.

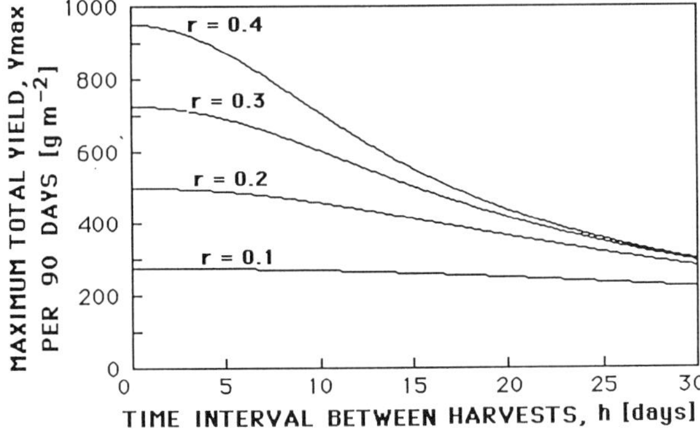

Figure 3. Maximum total yield obtainable in 90 days at four constant specific growth rates (r) W_{01} = 50 g m^{-2}, and K = 100 g m^{-2}, as a function of h, the fixed time interval between successive harvests.

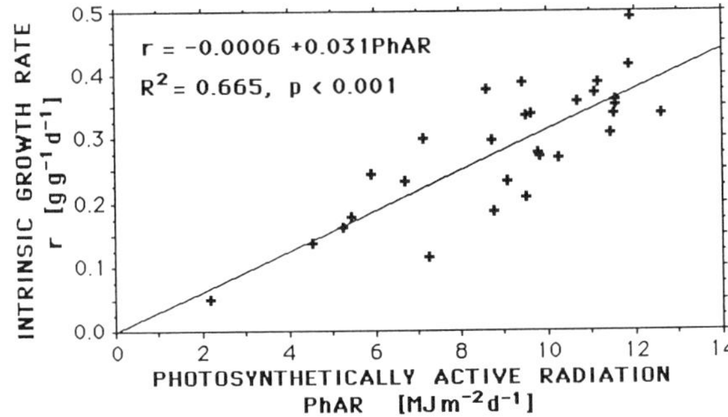

Figure 4. The dependence of giant duckweed (Spirodela polyrhiza) specific growth rate on daily PhAR; total (N) in water 10 mg l^{-1}; daily mean water temperature = 19.8°C (±2.5° C).

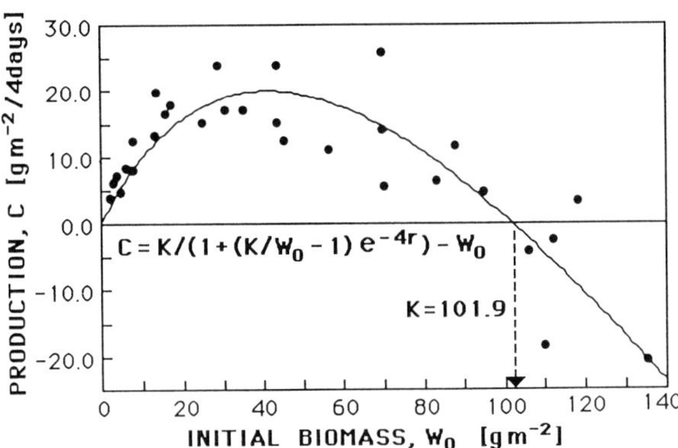

Figure 5. Experimental estimation of the carrying capacity, K, for giant duckweed (Spirodela polyrhiza) from the dependence of 4 days' biomass production, C, on the initial biomass, W_0.

frames (1 x 1 m) anchored in the bottom, which prevented the duckweed from being dispersed by wind (Fig. 7). During June and July 1982, O, 25, 50, 75% of duckweed biomass, each in four replicates, was harvested at intervals of three to five days for 35 days (Fig. 1). There was only a slight difference between maximum total yield under 25 and 50% harvest regimes: 163.9 and 157.7 g m^{-2}.

The intrinsic growth rate, was assessed in an independent experiment for duckweed with no crowding of individual plants; the regression of r on daily sums of incoming photosynthetically active radiation (PhAR) between 380 and 720 nm was calculated (Fig. 4).

The experiment to estimate carrying capacity for giant duckweed population was set up in a holding pond, State Fishery at Trebon, in July 1982, and repeated twice. Both sets of data were used to estimate the carrying capacity (K) by non-linear regression analysis (Fig. 5) at 101.9 g m^{-2} of dry mass. This is just the first approximation; K might be actually a slightly environment dependent parameter.

Simulation of giant duckweed (Spirodela polyrhiza) biomass growth with and without discrete harvests for the same period as the harvesting experiment (see Fig. 1) is shown in Figure 6. Two different optimal harvest fractions are calculated for each harvest date using the equation from Figure 4: (1) based on the actual PhAR in the following interval (solid line) and (2) based on PhAR in the previous interval (dashed line). Each harvest fraction is also dependent on the actual length of the following interval according to equation (3).

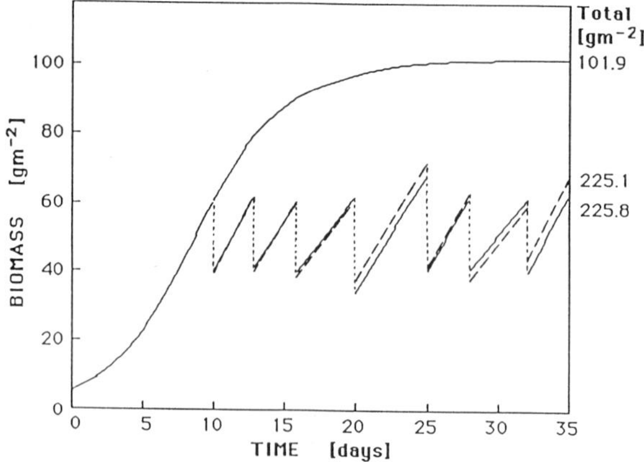

Figure 6. Simulation of giant duckweed (Spirodela polyrhiza) biomass growth with and without discrete harvests for the same period and the same intervals between harvests as in the harvesting experiments (see Fig. 1). Two different optimal harvest fractions, k, are calculated for each harvest date: on the basis of actual PhAR in the following interval (solid line) and on the basis of PhAR in the previous interval (dashed line) using the equation from Figure 4 for calculation of r. Each harvest fraction is also dependent on the actual length of the following interval. The simulated total yields are 101.9 (only final harvest), 225.8 (actual PhAR used) and 225.1 g m^{-2} (extrapolated PhAR used.)

Figure 7. Floating wooden frame (4 x 4 m; 1 m^2 each unit) for harvesting experiment with the giant duckweed (Spirodela polyrhiza).

The total yields are 101.9 (only final harvest), 225.8 (actual PhAR used), and 225.1 g m^{-2} (extrapolated PhAR used).

DISCUSSION AND CONCLUSIONS

The highest experimental yield obtained at 25% harvest fraction, 163.9 g m^{-2} of dry mass, falls short of the theoretically obtainable maximum yield by 27.2%. An important difference between experimental arbitrary harvests (Fig. 1) and simulated optimal harvests (Fig. 6) seems to be no harvest at the end of the first simulated interval: k calculated from (3) was negative. The average k_i in the subsequent simulated harvests was 0.36 (0.246 - 0.43), approximately half way between the two arbitrary harvest fractions (0.25 and 0.50, see Fig. 1). In view of the method used for calculating the optimum harvest fraction, it is obvious that the degree of accuracy of the estimate of the optimum harvest fraction should increase with the similarity of environmental conditions during the two successive intervals involved. There was, however, virtually no difference between the total simulated yield with optimal harvest fractions calculated on the basis of either the actual or extrapolated PhAR for each interval: 225.8 and 225.1 g m^{-2}, respectively, because the weather was fairly stable during the experiment.

The results presented should be regarded as a first approximation to the optimum harvesting under field conditions. The method can be improved in many ways. For example, the linear regression of r on PhAR used in the present study holds for PhAR < 13 MJ $m^{-2}d^{-1}$ provided the water temperature is about 20°C (±2.5°C). For a broader range of PhAR and temperature, a non-linear dependence is to be expected (see Wedge and Burris 1982; Rejmánková 1979). This will be important for implementation of the described method in warmer climates. A multiple regression of r on PhAR, nitrogen, and phosphorus concentrations would be desirable for situations with low and fluctuating concentrations of these elements.

ACKNOWLEDGEMENT

The experimental part of the presented research was conducted while the senior author worked as a research associate in the Department of Hydrobotany, Czechoslovak Academy of Sciences, at Trebon, Czechoslovakia. We are grateful to Dr. Karel Priban for providing us with the meterological data and to Dr. Alena Lukavska for chemical water analyses. Mr. F. Dedek constructed the wooden frames. Dr. D.D. Culley, who visted Trebon in 1981 and 1982, participated in numerous inspiring discussions on production and management of duckweed. Dr. G. W. Swan kindly reviewed the manuscript and gave us his valuable comments. We would also like to recognize the State Fishery at Trebon, namely Mr. Cermak for its cooperation. Dr. D.F. Spencer is to be thanked for reviewing the manuscript.

REFERENCES

Cassani, J.R., Caton, W.E. and Hansen, T.H. 1982. Culture and diet of hybrid grass carp fingerlings. Journal of Aquatic Plant Management 20:30-32.

Clatworthy, J.N. and Harper, J.H. 1962. Inter- and intraspecific interference within cultures of Lemna spp. and Salvinia natans. Journal Experimental Botany 13: 307-324.

Culley, D.D., Rejmankova, E., Frye, J.B. and Kvet, J. 1983. Production, chemical quality and use of duckweeds (Lemnaceae) in aquaculture, waste management and animal feeds. Proceedings of the World Maricultural Society 12: 27-49.

DeBusk, T.A., Ryther, J.H., Hanisak, M.D. and Williams, L.D. 1981. Effects of seasonality and plant density on the productivity of some freshwater macrophytes. Aquatic Botany 10: 133-142.

Elizarov, E.Ya. and Svirezev, Yu.M. 1970. Optimal productivity of biogeocenoses. Problemy Kibernetiky 22: 191-202. (In Russian).

Elizarov, E. Ya. and Svirezev, Yu. M. 1972. On optimal productivity of biosystems. Zhurnal Obshchey Biologii 33: 251-260. (In Russian).

Ericson, T., Larson, C. M. and Tillberg, E. 1982. Growth responses of Lemna to different levels of nitrogen limitation. Zeitschrift für Pflanzenphysiologie 105:331-340.

Ikusima, I. 1955. Growth of duckweed populations as related to frond density. Physiology and Ecology (Kyoto) 6: 69-81.

Landolt, E. 1987, The family of Lemnaceae - a monographic study. Volume 2, Veroeffentlichungen des Geobotanisches Institutes der ETH, Stiftung Rubel, Zurich 95:1-638.

Lorber, M.N., Mishoe, J.W. and Reddy, P.R. 1984. Modelling and analysis of water hyacinth biomass. Ecological Modelling 24:61-77.

Madsen, J.D., Adams, M.S. and Ruffier, P. 1988. Harvest as a control for sago pondweed (Potamogeton pectinatus L.) in Badfish creek, Wisconsin: Frequency, efficiency and its impact on the stream community oxygen. Journal of Aquatic Plant Management 26:20-25.

Mbagwu, I.G. and Adeniji, H.A. 1988. The nutritional content of duckweed (Lemna paucicostata Hegelm) in the Kainji Lake area, Nigeria. Aquatic Botany 29:357-366.

Mestayer, C.R., Culley, D.D., Standifer, L.C. and Koonce, K.L. 1984. Solar energy conversion efficiency and growth aspects of the duckweed, Spirodela punctata (G.F.W. Mey.) Thompson. Aquatic Botany 19:157-170.

Oron, G., Wildschut, L.R., and Porath, D. 1985. Wastewater recycling by duckweed for protein production and effluent renovation. Water Science and Technology 17: 803-817.

Painter, D.S. 1988. Long-term effects of mechanical harvesting on Eurasian watermilfoil. Journal of Aquatic Plant Management 26:25-29.

Porath, D. and Pollock, J. 1982. Ammonia stripping by duckweed and its feasibility in circulating aquaculture. Aquatic Botany 13: 125-131.

Reddy, K.R. 1983. Water hyacinth production system in nutrient rich waters. pp. 30-32. In W. Smith, ed. The Methane from Biomass and Waste Program Annual Report for 1982. Institute of Food and Agricultural Sciences, University of Florida, Gainesville, Fl. USA.

Reddy, K.R. and DeBusk, W.F. 1985a. Nutrient removal potential of selected aquatic macrophytes. Journal of Environmental Quality 14:459-462.

Reddy, K.R. and DeBusk, W.F. 1985b. Growth characteristics of aquatic macrophytes cultured in nutrient-enriched water. II. Azolla, duckweed, and Salvinia. Economic Botany 39:200-208.

Rejmánková, E. 1979. Function of duckweed in fishpond ecosystems. Ph.D. Dissertation, Institute of Botany. Czechoslovak Academy of Science, Praha , Czechoslovakia. 166p. (In Czech).

Rejmánková, E. 1982. The role of duckweeds (Lemnaceae) in small wetland water bodies of Czechoslovakia. pp. 397-403. In Gopal, B., Turner, R.E., Wetzel, R.G. and Wigham, D. F. eds. Wetlands: Ecology and Management. Proceedings of the First International Wetland Conference. International Scientific Publishers, Jaipur, India.

Seip, K.I. 1980. A computational model for growth and harvesting of the marine alga Ascophyllum nodosum. Ecological Modelling 8:189-199.

Swan, G.W. 1975. Some strategies for harvesting a single species, Bulletin Mathematical Biology 37:659-673.

Tucker, C.S. 1983. Culture density and productivity of Pistia stratiotes. Journal of Aquatic Plant Management, 21: 40-41.

Wedge, R.M. and Burris, J.E. 1982. Effect of light and temperature on duckweed photosynthesis. Aquatic Botany 13:133-140.

WETLAND RECLAMATION IN THE BYELORUSSIAN SSR.

D.I. Berezkin
Department of Hydrometerology and Environmental Control of the Byelorussian SSR
110 Lenin Avenue, SU - 220 838
Minsk, USSR

V.T. Klimkov
Byelorussian Institute for Amelioration and Water Management
Gorkij Street 153, SU - 220 040
Minsk, USSR

I.M. Nesterenko
Academy of Sciences of the Byelorussian SSR, Institute of Peat,
Staro-Boriosovsky 10, SU-220 030 Minsk, U.S.S.R.

J. Kvet
Institute of Botany, Czechoslovak Academy of Science,
CS-37982 Trebon, Czechoslovakia

INTRODUCTION

J. Kvet

The three short papers to follow describe wetland reclamation operations undertaken in the Byelorussian SSR (USSR) in the last few decades and some of their ecological consequences. The Byelorussian wetlands partly belong to a large wetland complex situated in the Polessye region which also encompasses the northern Ukraine and extends along the main European water divide, between the basins of the Black and the Baltic Sea. The river Pripyat, the largest tributary of the Dnieper river from the west, is the hydrological axis of the Polessye and receives most of its water discharge.

Small scale drainage operations were planned and accomplished here as early as the mid-nineteenth century, but a large scale scheme of wetland drainage and reclamation has been completed and put into operation since World War II. UNESCO's IHD (International Hydrological Decade) program held a conference on the "Reclamation of Marsh Ridden Areas" in Minsk, the capital of Byelorussia, in 1970 and the SCOPE (Scientific Committee on Problems of Environment of I.C.S.U.) project on "Ecosystem Dynamics in Continental Wetlands and Shallow Waters" organized one of its workshops in Byelorussia in 1981. Visits to some of the reclaimed as well as preserved wetland sites were on the program of both meetings. While the spirit of the 1970 conference was predominately optimistic and enthusiastically in favour of wetland reclamation, the 1981 workshop paid due attention to its both actual and potential unfavorable ecological consequences. Since then, this critical attitude has been widely acknowleged in the Soviet Union itself and has induced a re-assessment of land drainage and wetland reclamation, also because not all of its anticipated economic benefits did materialize: crop yields did not become as high as originally expected, high energy subsidies in the form of fertilizers, maintenance of drainage systems, occasional irrigation, etc. became necessary. On the other hand, enhanced peat mineralization, and subsidence became noticable and have reduced, in wet periods, the average relative fall of groundwater table brought about by wetland drainage.

The aim of the great part of recent investigations of reclaimed Byelorussian wetlands therefore is to find ways of preserving as much as possible of the benefits of wetland reclamation for agriculture and, at the same time, of minimizing its undesirable ecological consequences which also become economically unfavorable in long term run. This approach is evident from the three papers which are presented here as a small sample representing the large applied research effort spent on

D. F. Whigham et al. (eds.), Wetland Ecology and Management: Case Studies, 45–54.

problems of drainage and reclamation of wetlands (especially bogs), the results of which are contained in a voluminous Soviet literature (predominantly in the Russian Language) on this subject. (For some references see especially the paper by I.M. Nesterenko.)

Last but not least, one fact deserves to be acknowledged with satisfaction, namely that the planners of the wetland drainage and reclamation schemes in Byelorussia were more enlightened than their colleagues in many other countries: from the beginning, they set aside about 25% of the original wetlands, predominantly bogs situated at water divides, and parts of river or stream floodplains. Most of these areas have remained untouched and unmanaged, or only extensively managed by traditional methods (light grazing or mowing). These protected areas ("Zapovedniki") have preserved a good deal of the original biotic diversity and hydrological functions of the Byelorussian wetlands. Most famous is the Berezina "Zapovednik", situated to the north of Minsk (outside the Polessye region, near the water divide between the Dnieper and Northern Dvina River basins), which also became a Biosphere Reserve of UNESCO. The Byelorussian wetlands have thus at least partly retained their natural character of the largest freshwater wetlands in the European part of the U.S.S.R. But recent trends in U.S.S.R in both ecological and economic thinking and planning are promising signals also for the future of these precious wetlands.

EXCESSIVELY WET SOILS AND MICROCLIMATE

D. I. Berezkin

INTRODUCTION

Drainage of wetlands for land reclamation is a widespread process in the USSR; especially in areas such as the Byelorussian Polessye region where of the total area of 60,000 km^2, 27,000 km^2 are wetland areas (bogs, marshes and swamps). Many of the wetland habitats have a favorable growing season (205 - 210 days), have an adequate supply of water, and have highly fertile soils. These conditions are all good prerequisites for agricultural sites and there is strong pressure to drain them and convert them for agricultural development. Reclamation of wetland soils is aimed at providing optimum conditions for crop production.

Reclaimed wetland areas have a relatively high production potential but their agroclimatic and microclimatic features must also be taken into account. In this extended abstract, I describe research that has been directed toward an understanding of how much wetland reclamation has changed the water and heat regime and how these changes have affected crop growth and development.

The work reported here was primarily conducted in the Luninsk wetlands, which are located about 60 km from Pinsk. A primary objective was to address the question of the effects that reclamation has on surrounding plant and animal populations and on microclimate.

CHANGES ASSOCIATED WITH DRAINAGE

Drainage of wetlands substantially changes the hydrophysical characteristics of the soils. Drained peat has an increased heat capacity and a lower heat conductivity than undrained peat. Removal of wetland vegetation and subsequent cultivation results in a substantial decrease in surface roughness. Drainage and lowering the water table affect evaporation. Wind erosion from the dried and exposed peat surface increases the amount of atmospheric dust. Since the types of the changes just described depend especially on soil moisture content, presence of vegetation cover, degree of peat mineralization, and physical properties of peat, comparisons of wetland sites before and after reclamation are given.

Surface Temperature. The surface temperature of drained peat that was devoid of vegetation had a maximum that was 5-10°C higher and a minimum that was 2-5°C lower than those of undrained peat. The incidence of ground frosts increased and the minimum temperature reached 3 to 4°C even in summer months. A dense vegetation significantly altered all these temperature features in drained and unvegetated peatlands.

Temperature of the arable soil zone. Summer temperatures of the upper layer (20 cm) of soil in drained areas that were devoid of vegetation were 2 to 3°C lower than those in undrained bogs. Dried bogs that had a grass cover were 1 to 2°C cooler in the 0-20 cm depth interval than undrained bogs.

Air Temperature. Air temperature differences at night between drained and undrained wetlands were minimal at a height of 50 cm above the surface when there was a vegetation cover. When the grass cover was poorly developed, then air temperatures at night were usually cooler by 0.3-0.5°C over drained wetlands. (On clear nights the differences were as much as 2°C.) During the day there were no distinct air temperature differences between drained and undrained areas with and without a distinct vegetation cover. Frost was more severe by as much as 5 to 7°C at the height of 50 cm over a drained bog than over an undrained one.

Air Humidity. Differences in air humidity over drained and undrained bogs were quite variable but, in general, the number of dry days (daily minimum relative air humidity less than 30%) increased after drainage. For example, the number of dry days increased from three to six per year over drained wetlands in the Lunenetsk area.

Wind velocity. A decrease in surface roughness after reclamation led to an increase in wind velocity. In the drained Lunenetsk wetlands, the average wind velocity increased by 10% from 2.7 to 3.1 $m.s^{-1}$. At the same time, it did not change in areas where reclamation had not occurred.

Atmospheric Transparency. For the last 15 years, a general decrease in atmospheric transparancy has occurred because of increased air pollution. However, in the surroundings of the settlement of Polesskiy which has no industrial pollution, transparency has decreased more than in the industrial center of Minsk. In the Polesskiy area, transparency during the summer decreased by about 6% in 25 years. During the same 25 years, solar radiation input on clear and cloudy days decreased by 4% and 2%, respectively. The probable reason for the decrease in atmospheric transparency in the Polesskiy region was probably associated with increased dust due to wind erosion from reclaimed wetland areas.

Precipitation. No systematic changes in precipitation have been measured in areas that have been drained compared to those that have remained in a natural state.

GENERAL CONCLUSIONS

As a result of drainage and reclamation, the temperature regime changed at the soil surface and in the adjoining 10 cm layer of air. Daily maximum temperatures increased and minimum temperatures decreased. The incidence and intensity of frost also increased. Differences in air temperature decreased with height and practically none were found at 150 to 200 cm. The number of dry days increased while the mean relative air humidity changed little during the growing season. Mean wind velocity was greater over reclaimed wetlands. Atmospheric transparency decreased which resulted in a 2 to 4% decrease in total incoming radiation. No changes in the quantity of precipitation occurred over reclaimed areas, at least during the growing season.

Wetland reclamation may also have a great influence on rivers. Drainage, straightening and channelization of rivers, construction of large reservoirs and small ponds and water use for irrigation all influence the river ecosystem. Changes in the hydrographical regime are especially pronounced in small rivers that contain reclaimed wetlands.

Deepening of river-beds has resulted in increased erosion with the intensity and extent of land drainage. Water discharge in a drained landscape increased while less groundwater remained in the soil. The artificial hydrographical networks associated with large scale drainage also promoted run-off and resulted in an increase in stream discharge.

Numerous scientists and specialists have come to the conclusion that well-designed wetland reclamation projects have a favorable influence on river flow. It is possible, however, to speak only about the redistribution of run-off, and not the losses of water from the decreased water storage capacity of the reclaimed regions as a whole.

METHODS AND TECHNIQUES OF WETLAND RECLAMATION IN BYELORUSSIA

V.T. Klimkov

INTRODUCTION

In the Byelorussian region of the USSR, average precipitation exceeds average evaporation except that about once in every 5 years total evaporation between May and August exceeds precipitation by 100 - 200mm (Golchenko 1976). The climate, therefore, is quite ideal for agriculture. Because of the ideal conditions, many wetlands in the region have been drained and reclaimed for use in agricultural production.

Of 74,000 km^2 of land suitable for reclamation in the region, about 11,000 km^2 of peat/wetlands and 19,000 km^2 of wetlands on mineral soils have been reclaimed. Drainage and irrigation systems have been constructed over an area of about 6,000 km^2 and reclaimed areas account for approximately 20% of land used for agriculture. The reclaimed wetland areas yield about 30% of the total crop production (Anonymous 1979b).

Depending on water supply, hydrogeological conditions and the aims of reclamation, several methods are used to control water tables: 1) increase the rate at which surface water is removed, 2) prevent floods, and 3) improve hydrologic and physical properties of soils. The purpose of this paper is to describe the types of wetlands that are reclaimed and describe the methods used to reclaim them, either singly or in combination.

TYPES OF WETLANDS THAT ARE RECLAIMED

Wetlands with Peat Substrates

Peat bogs occupy more than 29,000 km^2 and occur in all regions of the Byelorussian Republic. Peat deposits are 1 m or more thick in 46% of bogs (Anoshko 1978). In lowland bogs that are underlain by loam and clay loam, and have gravity groundwater flow, and in bogs with horizontal interval groundwater flow, blind drainage channels are used; they are dug parallel to each other and are 20 to 40 m apart. In lowland bogs that are underlain with thick highly permeable sands and where water conductivity is > 10 meters per day, 1.6 to 1.8 m deep open drains are cut into the subsoil. Drainage is efficient when the drains are spaced at 400 m. When water conductivity is between 1 and 10 $m.d^{-1}$ a drainage system employing widely spaced open drains is necessary. In recent years vertically cut ditches with large diameter pipes have been used under such conditions in order to reduce the depth of open drains.

Bogs with a peat layer more than 1.5 m thick are drained in two stages: First, a network of open drains is dug and the surface peat is excavated. Three to five years later open drains are replaced by closed ones. At the same time, the surface of the reclaimed areas is levelled and additional measures are taken to improve soil fertility.

A more recent reclamation technique is employed on sites with highly permeable subsoils. Vertical drainage pipes are used to control the moisture content in the rhizosphere. In wet periods the system drains while during dry seasons the moisture content of the soil is increased by using groundwater for spray irrigation. Groundwater used in the dry periods is then replenished during wet periods. Vertical drainage systems can be applied only to areas larger than 10,000 km^2.

Wetlands on Mineral Soils

Waterlogged mineral soils cover 45.000 km^2 and include 21,000 km^2 of heavy soils (Brusilovskii 1981). For non-cohesive soils, systematic horizontal drainage is most efficient. In heavy soils, drainage alone does not ensure the required removal of excess water, and additional measures are used to improve surface runoff and increase the water-holding capacity and conductivity of the soil. A network of intercepting channels is constructed, depressions are filled with soil, and the ground surface is levelled. If depressions with no outlets are not filled they are turned into drain sumps. In some cases, artificial hollows are made to let out water from the depressions. The most efficient measure for improving the water-holding capacity of heavy soils is to loosen the soil to a depth of 0.6 m along with an enhancement of surface runoff by systematic drainage.

Polders

Polder construction began in the 1960's on floodplains of rivers, and around lakes and shallow reservoirs. Non-irrigated polders now occupy more than 1000 km^2. Irrigated polders cover a much smaller area and have been developed in areas that flood for more than 100 days each year and they are only suitable for hay production. Water recycling systems are used in most polders to minimize eutrophication of areas that receive polder water and to store water for later use with spray irrigation systems.

TECHNIQUES FOR THE RECLAMATION OF WETLANDS

Open Drains

Open ditches are used to drain forested bogs and mires with thin peat layers. Open drains are also used to separate reclaimed areas from external water sources, for pre-drainage of bogs, and as water intakes of drainage systems. They usually have a trapezoidal shape in cross-section and the slope of the ditch wall depends on soil type. The bottoms of the ditches are also sloped to prevent the growth of plants and to minimize erosion. Where water flow is rapid, drain beds are consolidated with sod, brushwood, gravel, crushed rock, and reinforced concrete slabs.

Closed Drains

Byelorussia has almost 130-years of experience with closed drainage systems. They found their widest application after World War II and they are now used in more than 14,000

km^2 of drained wetland area. Usually, the spacing of drains is adjusted to the local conditions and drain depth is 0.8 - 1.2 m. Drains are deeper on peatlands to match the estimated peat settling by the end of the drain life.

Vertical Drains

Vertical drainage is achieved by using 20 to 40 m deep water wells. This is the most efficient drainage and irrigation system because it eliminates the need for reservoirs. If the wells are not too widely spaced they are often connected by horizontal underground siphon pipes. Vertical drainage systems are most efficient if the aquifer thickness is at least 15 m and the water conductivity is more than 150 m. $^{-1}$ (Murashko 1980). Also, the topsoil must be highly permeable and hydraulically connected to the aquifer. For uniform drainage in systems with vertical drains, the slope of the area must be less than 0.5 m.

Control of Water Regime

Depending on edaphic and climatic conditions and on the agricultural use of the land, various combinations of the measures described are employed to ensure favourable soil moisture conditions for crop sowing, cultivation, and harvest on reclaimed wet soils.

In areas with a high or medium subsoil permeability, irrigation is performed by capillary feeding from a controlled groundwater table that is maintained by pumping water into open and/or closed drain systems. Water needed to maintain a suitable groundwater level is stored in pools and reservoirs during wet periods. These reservoirs are often used for many purposes such as water supply, fishing and irrigation (Anonymous 1977). In other areas, adequate soil moisture is maintained by spraying at rates of 0.2, 0.3, and 0.8 mm. min^{-1} on heavy, medium, and porous soils, respectively. Irrigation rates are reduced on sloping sites. In dry years, 3 to 8 sprays are required during the growing season and during most years irrigated areas receive 180 to 225 mm of water. Stationary and mobile sprinklers may be either of the rotary or boom type.

Crop production on the reclaimed wetland areas varies from 3.4 to 5.5 t · ha^{-1} for grain crops, between 12 and 16 t · ha^{-1} for hay, and between 30 and 50 t · ha^{-1} for potatoes (Anonymous 1980).

CHANGES IN THE HYDROPHYSICAL PROPERTIES OF PEAT SOIL FOLLOWING THEIR DRAINAGE

I.M. Nesterenko

INTRODUCTION

Natural bogs may be regarded as both solar "storages" and rich scientific "archives" in which nature has accumulated valuable data on the formation and development of peat soils for thousands of years. Mire and peat deposits have been characterized in many countries (Kivinen 1980; Anonymous 1976) and the conditions

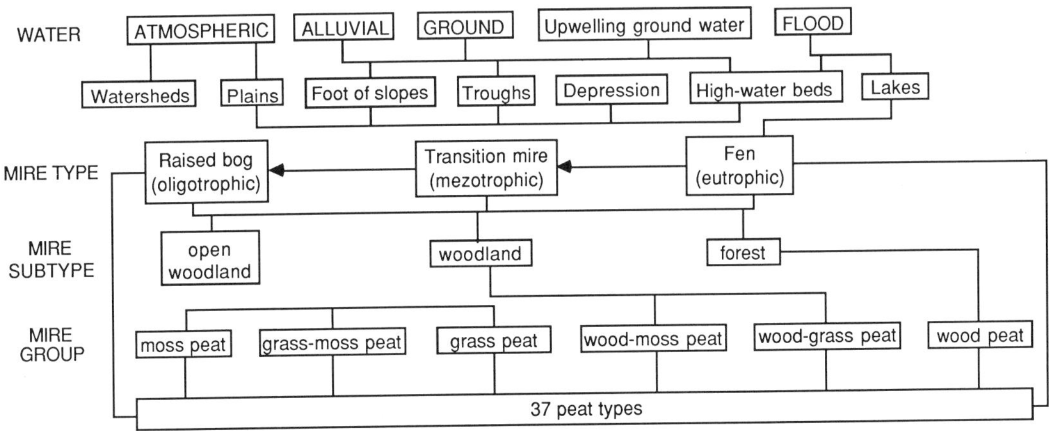

Figure 1. Conditions of peat formation and classification and mire types according to the proposals of the International Peat Society (Anonymous 1976; Kivinen 1980).

Table 1. Percent decomposition (R) and percent ash content (A) of peats in Byelorussia and Karelia. Values are means ± 1 standard deviation. Sample sizes are given in parenthesis.

Raised-bog Peat		Transitional Peat		Fen Peat	
R	A	R	A	R	A
		Byelorussia			
23 ± 16.7 (785)	2.4 ± 1.3 (785)	31 ± 11.5 (321)	4.7 ± 2.56 (321)	34 ± 11.7 (908)	7.6 ± 3.1 (908)
		Karelia			
26.3 ± 9.33 (111	3.53 ± 1.63 (97)	31.7 ± 8.09 (286)	5.09 ± 2.73 (246)	31.1 ± 7.79 (205)	9.57 ± 6.78 (163)

under which mires are formed (hydrologic and geomorphic) are well known and their important physical characteristics have been studied (Fig. 1). A large volume of information has now accumulated on the genesis and dynamics of bogs, the nature of peat as a geological formation, its technical and hydrological properties, elementary and group chemical composition, and the chemical compositions of ash and other characteristics (Gerasimov 1930; Ivanov 1975; Kac 1948; Lishtvan and Kord 1975; Neishtadt 1957).

Interest in mires has grown because of their potential uses. Peat is a valuable natural resource. Mires cover above $3.4 \cdot 10^6$ km^2 and forecasts show that the figure is likely to increase to $5 \cdot 10^6$ km^2 (Kivinen 1981).

In recent years, there has been an increased rate of reclamation of mires for agricultural and forestry practices. For agricultural purposes, the availability of nutrients for plant growth is important and nutrients often need to be applied. One consequence of fertilization and lime application is that mineralization rates are lower for 5-10 years following drainage (Havelka and Haken 1969; Lishtvan and Korol 1975). In general, however, the agrochemical properties of cultivated peat soils vary little between mire types and origins, and depend on the technological properties of the peat. The regulation of water and the oxidation-reduction status of peat soil is much more important and many techniques are applied in various types of reclaimed peat areas. Drainage channels and other facilities are designed according to conventional empirical and statistical indices of the peat. The most important information is the precipitation and evaporation regime of the area to be reclaimed, modulus of flow, drainage rate, and time when excess water needs to be removed.

In this paper, I describe the physical characteristics of peat soils in their natural and drained conditions. The information presented in this paper has been very useful in the USSR where many wetlands are drained for a variety of reasons.

CHARACTERISTICS OF PEAT

Decomposition and Ash Content

The degree of decomposition (R), botanical composition, and ash content (A) are good indices of peat properties (Lishtvan and Korol 1975) and the statistical relationships of R and A for peats in Soviet Karelia are given in Table 1 (Nesterenko 1979). The characteristics of peat in Karelia are similar to those in Byelorussia. The degree of decomposition and ash content are normally distributed in both areas and there is no difference in R (Table1). Differences in the degree of decomposition are more closely related to different peat groups. Decomposition rates are almost identical for transition and fen peat bogs and no differences have been found between the same peat types in different geographic zones. The similarities may, however, be explained by the disadvantages of the visual technique used to assess R (normal

Table 2. Ranges in the degree of decomposition (R), ash content (A) and pH of peat from the three main mire types and peat groups (Lishtvan and Korol 1975).

Mire Types	R	A	pH
Fen	21 - 45	5.6 - 9.6	4.8 - 5.3
Transition	22 - 45	3.5 - 7.1	4.0 - 4.6
Raised-bog	13 - 55	2.3 - 3.8	3.2 - 4.6
Peat Groups			
Woody	45 - 55	3.8 - 9.6	4.6 - 5.3
Woody-grass	39 - 51	2.8 - 8.0	3.2 - 5.0
Woody-moss	35 - 38	3.5 - 7.8	3.5 - 5.0
Grass	29 - 37	2.6 - 6.7	3.5 - 5.0
Grass-moss	24 - 30	2.3 - 5.6	3.3 - 4.8
Moss	13 - 21	2.3 - 6.5	3.2 - 4.9

accuracy 5%), by averaging R for a whole peat deposit, by the same plant species occurring in various mires, and by the preservation of most decomposition resistant species in peat deposits.

Differences in ash content are considerable within mire types and peat groups (Table 2) and are primarily related to variation in water supply and mineral content in different parts of the mires during their formation.

Regression coefficients for all samples that have been analyzed are fairly similar (0.74-0.85 range) and indicate that a change in ash content of 1% results in a density change between 0.011 and 0.015 g cm^{-3}. The intercept of the regression between ash content and density (1.45-1.51) indicates the minimum density for the organic matter of peat. This points to a physical link existing between the above characteristics.

For a wide variety of peat samples in the USSR, the relationship between density (r) and ash content is shown in the following equation:

$$r = 0.013\,A + 1.48 \qquad \text{(Eq. 1)}$$

These characteristics of organic soils have been corroborated by numerous studies.

Ash content varies between 1 and 8% (Lishtvan and Korol 1975; Nesterenko 1976 and 1979; Okruszko 1971). The density and porosity of peat varies little and peat has a high water holding capacity (86 to 94%). According to Lishtvan and Korol (1975), the content of physiochemically bound water is practically the same in raised-bog and fen peats (average 46-50%). This is a common feature and the major prerequisite for the formation and development of mires. The presence of a biologically active top peat layer (0.3-0.4 m) deep and a denser deeper layer differing in capillary and filtering properties insure the high self-regulating capacity of mire systems. When the groundwater level (GWL) drops to a lower boundary of the active layer, both discharge and evaporation are reduced considerably and water in the deeper layer remains practically unaltered. When precipitation occurs, the GWL rises rapidly to recharge the active peat layer and provides stable environmental conditions for the vegetation, for the maintenance of a near-optimum water regime, and for stability of the peat ecosystem.

Bulk Density

Bulk density (γ) is one of the major characteristics of peat soils because it controls the hydrophysical characteristics needed to estimate the organic matter and mineral nutrient content and to forecast possible changes in properties of the peat following reclamation. Numerous studies have indicated that substantial changes in γ occur in the top

Table 3. Mean bulk density (g cm^{-3}) of drained peat soils in the Korza lowland, (southern Karelia) between 1962 and 1982.

Depth (cm)	Bulk Density Prior to Drainage	1962	1966	1970	1977	1983
0 - 20	.08	.19	.21	.23	.25	.23
20 - 50	.10	.15	.16	.16	.18	.18
50 -100	.11	.11	.12	.13	.14	.14

layer of a peat in the first 1-2 years following drainage. These changes are due primarily to alteration of the top layer of peat during reclamation when it is mixed with underlying peat. Furthermore, it is during the initial period of reclamation that the most intense shrinkage and compression of the peat occurs. Bulk density changes little subsequently if reclaimed mires are used for growing perennial grasses. This conclusion is supported by long-term observations on mires in southern Karelia (Table 3).

Compression was obvious in the 0-50 cm depth interval a few years following drainage and during a dry period between 1972 and 1975 when ground-water levels fell to between 1.2 and 1.5 m. The stability of peat in the 50-100 cm depth interval provides evidence that peat compression gradually decreases with depth in reclaimed areas.

The variation of γ with time (T) in reclaimed peat soils was expressed in general terms by Maslov (1970) as:

$$\gamma = \gamma_0 \cdot (1 + mT^n) \qquad \text{(Eq. 2)}$$

where γ_o is the initial bulk density and T is time in years. Observations conducted for over 40 years at the Minsk Mire Experimental Station (Zubets and Dubrova 1974) have led to the formulation of the following equation to calculate bulk density in a layer of 0 to 30 cm thickness:

$$\gamma = \gamma_0 + 0.034\sqrt{T} \qquad \text{(Eq. 3)}$$

In different areas, γ of the top layers of drained peat ranges from (0.2 to 0.3 $g \cdot cm^{-3}$) between 20 and 100 years after reclamation. In deeper peat layers, γ varied between 0.1 and 0.2 $g \cdot cm^{-3}$.

Porosity and Water-holding Capacity

Porosity (P) and, consequently, the water holding capacity (W_{mw}) are related to bulk density (Maslov 1972 ; Anonymous 1976 b; Nesterenko 1976, 1979; Okruszko 1971). The relationship can be expressed as:

$$P = 95.0 - 34.8\,\gamma \qquad \text{(Eq. 4)}$$

The equation shows that a 3.5% decline in water holding-capacity would occur with an increase in the bulk density of 0.1 $g \cdot cm^{-3}$. The intercept (95.0) represents the maximum water-holding capacity of low-ash raised-bog peat.

This relationship is valid for both peat and mineral soils. The dependence given reflects both directed changes in hydrological properties of soils correlated with alterations of bulk density. It also proves the validity of the "natural" relationship for the entire ecological soil series.

The difference between the maximum amount of water a substrate can hold (W_{mw}) and field capacity (W_{fm}) determines the maximum water yield ($\mathcal{S}_{max}$).

At any height H (cm) above the ground water table (GWL) up to that of capillary rise (Hc), provided a balanced water distribution, represents a straight line, the specific water ($\mathcal{S}_H$) yield is:

$$\mathcal{S}_H = \frac{H}{H_c} \cdot \mathcal{S}_{max} \qquad \text{(Eq. 5)}$$

and the average specific yield from the estimated layer above GWL is

$$\mathcal{S} = \frac{H}{2H_c} \cdot \mathcal{S}_{max} \qquad \text{(Eq. 6)}$$

The water reserve W (in mm) above GWL in homogeneous soils for the layer showing H $\leq$ Hc is:

$$W_H = \frac{H}{10}\left(W_{mw} - \frac{H}{2H_c} \cdot \mathcal{S}_{max}\right) \qquad \text{(Eq.7)}$$

With average W_{mw} values of 85%, Hc - 100 cm

With average W_{mw} values of 85%, Hc - 100 cm

and

$\mathcal{S}_{max}$ = 15%. W_H can then be calculated as

$$W_H = 8.5\,H - 0.0075H^2 \qquad \text{(Eq. 8)}$$

Regression analysis of data obtained from observations of the water regime in a 0 to 1 m thick layer (Lorza lowland, southern Karelia) [W_H = 8.7 H - 0.00743 H^2 - 53; r = 0.91] revealed a relationship close to the one given in Equation 8.

Practically, in peat soils the water reserve in the layer above GWL is directly proportional to the depth at which the groundwater occurs and is 7.44 and 7.32 H in Karelia and the Leningrad region, respectively. The optimum depth, at which GWL occurs (drainage rate) in the USSR, provided perennial grasses are grown on peatlands, is recommended to be within 0.6 to 0.7 m in the first month of vegetation and 0.7 to 0.9 m by the end of the vegetation season (Lishtvan and Korol 1975). In the German Democratic Republic, Czechoslovakia, and the USA, higher groundwater levels, namely 0.4 to 0.6 m are recommended (Burks and Hare 1962; Lene 1972). Provisions for optimum conditions for crops (chiefly perennial grasses) create a stable ecological system in drained mires. To remain stable, it is particularly necessary to provide a stable moisture regime which, in dry areas, requires water reserves such as the large-scale measures that have been taken in the Ukrainian and Byelorussian Polessye region (Anonymous 1979a, 1982).

Subsidence

By maintaining high groundwater levels peat can be preserved longer and numerous studies have shown that the subsidence and decomposition of peat depend chiefly on drainage depth (h in meters), the initial peat depth (H_0 in meters) and the duration of drainage (T in years). Murashko (1981) has proposed the following formulas to calculate surface and subsurface subsidence:

$$\Delta H = AH_0 \{ 1 - \exp [-h(a+bt)] \} \quad \text{(Eq. 9)}$$

$$\Delta H = A(H_0 - h)\{1 - \exp [-h(c + dt)]\} \quad \text{(Eq. 10)}$$

where A is the peat density coefficient which is dependent either on bulk density or on humidity and degree of peat decomposition, a and c are the compression coefficients in the first year of drainage and b and d are those for subsequent years. Values for these variables in the European USSR are: $a = 0.07\ m^{-1}$; $b = 0.006\ m^{-1} . year^{-1}$; $c = 0.0021\ m^{-1}$; $d - 0.005\ m^{-1} . year^{-1}$.

Analysis of long-term data has revealed simple empirical relationships for peat subsidence (ΔH, in cm) for two areas of the Russian Federation (Chernenok 1966):

$$\Delta H = 11.9 \sqrt{T} \text{ for h between 1.7 and 2.0 m} \quad \text{(Eq. 11)}$$

and

$$\Delta H = 6.7 \sqrt{T} \text{ for h between 0.8 and 1.2 m} \quad \text{(Eq. 12)}$$

For the European North of the Russian Federation (Nesterenko 1979), the relationship is:

$$\Delta H = 7.15 \quad \eta\ h - T \quad \text{(Eq. 13)}$$

where the value of the correction coefficient (η) is dependent on the initial thickness of peat deposit as shown in Table 4. A theoretically probable value for peat subsidence of an estimated layer with a thickness H_0 (without taking into account the inflow and outflow of materials) is:

$$\Delta H = H_0 - H_1 = H_1 \left(1 - \frac{\gamma_0}{\gamma_1}\right) \quad \text{(Eq. 14)}$$

The value for a decomposing peat layer (H_p) can be calculated from the equation:

$$H_p = H_0 - H_1 \quad \frac{\gamma_1}{\gamma_0} + \frac{m_s}{100\gamma_0} \quad \text{(Eq. 15)}$$

Where H_0 and H_1 are the initial and subsequent peat thickness (cm), respectively; and γ_0 and γ_1 are the initial and subsequent bulk densities (g cm^{-3}); m_s is the balance of mineral substances brought in with fertilizers, atmospheric precipitation and lost with crop yield and groundwater discharge (t·ha^{-1}). This equation shows that in drained mires, bulk density should be one of the principal characteristics together with geobotanical composition, ash content, and some closely related water-physical characteristics such as coefficients of permeability, water yield, water capacity, water and heat conductivity.

Table 4. Correction coefficients η in eq. (19) for different values of the initial thickness of a peat deposit.

	Initial Peat Thickness				
H_0	4.0	3.0	2.0	1.0	0.5
η	1.0	0.8	0.55	0.3	0.2

In Karelia, peat subsidence does not exceed 5 mm· year^{-1} 20 years after drainage. The subsidence results chiefly from the mechanical packing of peat. However, considerable primary post-construction subsidence and a more intense compression of peat in the first post-drainage years (Nesterenko 1979) brought about a sink of the surface by 0.7 to 0.8 m in the Korza fen and the drain depth decreased to 0.8 - 1.0 m. In this area, the groundwater feeding rate is intense (up to 150 mm. year^{-1}), and in years with abundant precipitation wooden drainage does not provide the required control of the water regime. It is then necessary to reconstruct the network. At this second stage, alterations in the hydrophysical properties of peat are usually slower.

REFERENCES

Anonymous. 1976a. Transactions of the working group for peat classification. Helsinki.

Anonymous 1976b. Agrophysical properties of soils in the non-chernozem zone of the European part of the USSR. Moscow, Kolor, 368 p. (In Russian).

Anonymous. 1977. Development, irrigation and use of cultivated pastures. Minsk. USSR. (In Russian)

Anonymous. 1979a. Land amelioration in the Polessye and Environmental Protection. Report of the Ukrainian Res. Inst. for hydrotechnics and amelioration. Kiev, USSR. 83 p. (In Russian).

Anonymous. 1979b Rational soil use in reclaimed peat-bogs. Minsk. USSR. (In Russian)

Anonymous. 1980. Complex control of habitat conditions for crops in peaty soils. Minsk. USSR. (In Russian)

Anonymous. 1982. Problems of the Polessye Region no. 8. Nauka i Technika, Minsk, USSR. 280 p. (In Russian).

Anoshko, V.S. 1978. Geographical aspects of land reclamation in Byelorussia. Minsk. USSR. (In Russian)

Brusilovskii, Sh.I. 1981. Reclamation of heavy mineral soils. Minsk. USSR. (In Russian)

Burks, W. and P.J. Hare. 1962. Problems of peatland. Wellington. New Zealand.

Chernenok, V. Ya. 1966. Deep closed drainage for bog reclamation. Gidrotekhnika i Melioratsia 11: 19-24. (In Russian).

Gerasimov, D.A. 1930. Genetic classification of peat types. Torfyanoe delo No. 12, Moscow. USSR. (In Russian).

Golchenko, M.G. 1976. Provisions for a favorable soil moisture content and irrigation in Byelorussia. Minsk. USSR. (In Russian)

Havelka, F. and D. Haken. 1969. Soil-formation processes in cultivated bogs. In: Changes in peat soils after drainage and cultivation, pp. 173-181. Minsk. USSR. (In Russian).

Ivanov, K.E. 1975. Water budget in peat regions. Leningrad, USSR. 280 p. (In Russian).

Kac, N.L. 1948. Bog types of the USSR and Western Europe and their geographical distribution. Moscow, USSR. 320 p. (In Russian).

Kivinen, E. 1980. Proposal for a general classification of virgin peat. Proceedings 6th Intern. Peat Congress, Commission 1 - Duluth, Minnesota, U.S.A. 1980.

Kivinen, E. 1981. Utilization of peatland in some countries. Bulletin of the Intern. Peat Society 12:21-27.

Lene, P. 1972. Durch Anwendung neuer Verfahren der Grundwasser-Regulierung zu höheren Erträgen auf Niederungs-standorten. Berlin, Feldwirtschaft.

Lishtvan, I.I. and N.T. Korol. 1975. Essential properties of peat and methods of their assessment. Minsk, Nauka i Tekhnika, 319 p. (In Russian).

Maslov, B.S. 1970. Groundwater regime in wetlands and its control. Moscow, USSR. Kolos, 232 p. (In Russian).

Maslov, B.S. 1972. The relationship between the porosity and bulk density of drained soil. Express-Information Bulletin of the Information Center of the Ministry of Amelioration of the USSR, Moscow, ser. 2(4):3-5.

Murashko, A.I. 1980. Land drainage employing vertical drains. Minsk. USSR. (In Russian)

Neishtadt, M.I. 1957. History of forests and paleogeography of the U.S.S.R. Holocene period. Moscow, USSR. 403 p. (In Russian).

Nesterenko, I.M. 1976. Subsidence and wearing out of peat soils as a result of reclamation and agricultural utilization of marshlands. In Proceedings 5th Intern. Peat Congress Vol. 1, pp. 219-232, Poznan, Poland.

Nesterenko, I.M. 1979. Land amelioration in the northern part of the European USSR. Leningrad, Nauka, 356 p.

Okruszko, H. 1971. Determination of bulk density in hydrogenic soils based on their mineral content. Wiadomosci Instytutu Melioracji i Uzytkov zielonych, Warszawa 10(1):47-54. (In Polish).

Zubets, V.M. and Dubrova, V.I. 1974. Drainage and hydrophysical properties of peat. In: Amelioration and problems of organic matter, pp. 29-43, Minsk. USSR. (In Russian).

EFFECTS OF HUMAN ACTIVITIES AND SEA LEVEL RISE ON WETLAND ECOSYSTEMS IN THE CAPE FEAR RIVER ESTUARY, NORTH CAROLINA, USA

Courtney T. Hackney
Department of Biological Sciences
University of North Carolina at Wilmington
Wilmington, NC 28403

and

G. Frank Yelverton
Environmental Resources Branch
U.S. Army Corps of Engineers
Wilmington District
Wilmington, NC 28402

ABSTRACT

There have been tremendous changes in wetlands bordering the Cape Fear estuary since the first permanent European settlement in the 1720's. First, tidal freshwater swamps were cleared of cypress (Taxodium distichum) and gum (Nyssa sylvatica), diked, managed for rice production and finally abandoned around 1908. During the past 100 years the main channel through the estuary has been widened and deepened, increasing the tidal range at Wilmington, North Carolina by 53 cm. This produced a high tide 26.5 cm above previous levels. Also during the last century relative sea level has risen 30 cm. Today these increases result in the average high tide being 56.5 cm higher than in 1885. Wetlands inhabiting abandoned rice fields and swamps previously not subject to tidal flooding and/or to salinities above that of fresh water are now being affected. The intrusion of saline water into these wetlands has led to the conversion of freshwater tidal swamps into brackish marshes in the middle reaches of the estuary. Swamps in the upper reaches are gradually disintegrating and being replaced by oligohaline tidal marshes. Tropical and extratropical storms have the potential for pushing more saline water farther into the estuary and may accelerate conversion of swamp to marsh.

INTRODUCTION

Natural changes in wetland communities are generally associated with ecological succession: the wetland community is just one stage in the sere. Termed hydrarch succession because of the aquatic nature of the pioneer stages, this successional sequence leads to an upland type climax community (Oosting 1956). In general, wetlands accumulate materials and elevate their soil levels through either autochthonous material accumulation, mostly organic, or by allochthonous inputs of riverine or marine origin (Oosting 1956; Redfield 1972).

Human activities can dramatically alter the time frame of the successional sequence by altering sediment inputs. Dredging, dam construction, and jetties to protect harbors all may directly or indirectly alter nearby wetlands. Coastal wetlands are also subject to a number of other natural forces. Storm surges force saline water far up rivers damaging salt sensitive plant species, while sea level rise may produce similar results more slowly.

This paper examines physical changes in the Cape Fear River estuary, North Carolina, and documents concurrent changes in wetland communities bordering the estuary.

MATERIALS AND METHODS

Aerial photographs used in this study were obtained from the Wilmington, NC office of the Soil Conservation Service and from the Wilmington District office of the U.S. Army Corps of Engineers. Tidal data, including data on sea level rise were obtained from the

D. F. Whigham et al. (eds.), Wetland Ecology and Management: Case Studies, 55–61.

National Oceanic and Atmospheric Administration in Rockville, MD. Elevation of the marsh surface was measured by standard survey techniques and related to nearby benchmarks. Organic content of the soil was measured by ashing dried samples at 600°C for 6 h and salinity by the silver nitrate technique (Welcher 1963). Interstitial water samples were collected following the technique of Hackney and de la Cruz (1978), which consisted of removing soil cores and analyzing water which seeped from the soil into the resultant hole.

THE CAPE FEAR ESTUARY

Agricultural Alterations

When Europeans arrived in the Cape Fear Region of North Carolina they found an environment changed little by the amerindian inhabitants. After the first permanent settlement in the region in 1720, land adjoining the Cape Fear River and its tributary, the Northeast Cape Fear, was divided into plantations. Rice culture began soon afterwards and much of the inland (non-tidal) swamps were cleared and diked for growing rice. Following the American Revolution (1776) growers learned to harness tides to irrigate their rice fields and as a consequence much of the tidal wetlands along the River were also converted to rice culture. Conversion was accomplished by building a large dike along the river, thus isolating the tidal wetlands. Cypress (Taxodium distichum) and gum (Nyssa sylvatica) were cleared from the enclosed area and the land divided by smaller dikes into approximately 20 acre (8 ha) fields. Each field was drained by shallow ditches which further divided the land into 1 acre (0.4 ha) plots. Shallow ditches were designed to allow water to flow rapidly on and off the rice. These shallow ditches connected with larger canals which reached to the river. Flow to shallow ditches was controlled at the larger canals by swinging floodgates which opened and closed with the tides. Larger canals connected to the river were controlled by gates which would be opened or closed depending on the stage of rice growth. This system was successful until 1865 when the Civil War disrupted the area. It was practiced in a reduced fashion until 1908 after which it was abandoned. The previous summary is from Clifton (1973).

Rice is not salt tolerant and would not grow if fields were flooded with even slightly saline water. Saline (salinity > 1 o/oo) water regularly reaches well into the Northeast Cape Fear River today flooding all of the old rice fields (Fig. 1). Advertisements for land along the Cape Fear River in the late 1700's stated that tides flooding these wetlands consisted of fresh water only (Rogers 1970). Also, healthy cypress and gum swamps are found today only in those portions of the upper estuary not subject to salt incursions. Such communities were described as prime sites for construction of rice fields in the 1700's (Clifton 1973). If tidal marshlands were used they must certainly have been flooded by fresh water as draining marshes regularly flooded with saline waters produces very acidic soil conditions (Waisel 1972). Thus, location of tidal rice fields is a reliable indicator of the previous extent of freshwater in the Cape Fear Estuary (Fig. 1).

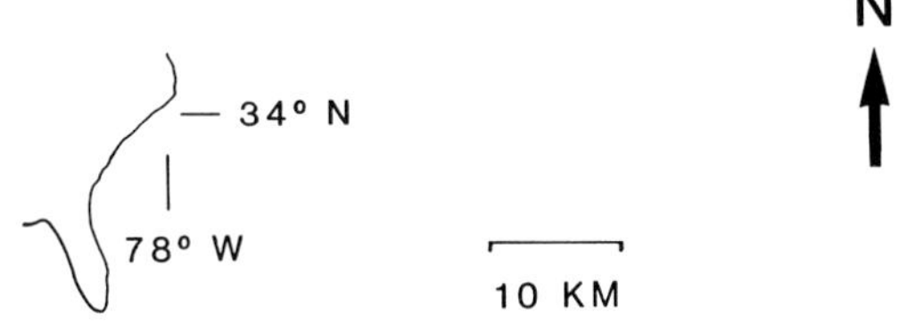

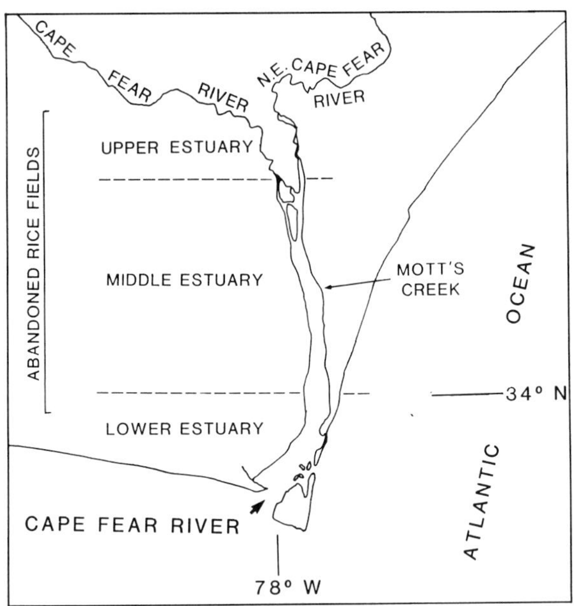

Figure 1. Cape Fear River, North Carolina showing the two major tributaries flowing into the estuary. The tidal surge actually extends into the tributaries north of the area shown. The upper estuary is primarily oligohaline in nature. Extensive tidal swamps are found along the Cape Fear River and the Northeast Cape Fear River north of their confluence.

River Dredging

Early in the history of North Carolina the Cape Fear River became an important transportation artery and was maintained for shipping through dredging and snag removal. Since 1889, the river channel has been dredged from 4.9 m deep and 82.3 m wide to todays depth of 11.6 m and a width of 122 m (U.S. Army Corps of Engineers 1977). Besides allowing longer, deeper draft vessels into Wilmington Harbor these changes allow a greater volume of water to be carried upstream with each flood tide.

According to the earliest tidal records (non-continuous) maintained by the National Oceanic Survey, the tidal range at Wilmington averaged 78 cm between 1908 and 1911. This followed dredging of the harbor to 6.1 m deep by 82.3 m wide which was completed in 1907. However, from 1972 (after the last major dredging operation in the river) to 1984 the tidal range averaged 130 cm (Fig. 2). This is a 53 cm increase in tidal range at Wilmington and a net increase of 26.5 cm in the relative height of the average tide. Continuous tidal gauge records maintained by the National Oceanographic Survey at Wilmington since 1936 also reflect this increase in tidal range. From 1936 to 1984 a 26 cm increase (104 to 130 cm) in tidal range has occurred which is associated with major dredging activities in the river (Fig. 2). The rise was not steady due to short-term, year to year change in astronomical conditions and long-term (19 year) National Tidal Datum Epoch (NOAA 1985).

While the tidal range at Wilmington has significantly changed during the last century, the tidal range at the mouth of the river (131 cm) has not changed since the first record in 1912 (NOAA 1985). Therefore, dredging has changed the tidal amplitude at Wilmington, equating it with the range at the river mouth 44 km downstream.

Sea Level Rise

National Ocean Records (NOAA 1985) also show a steady rise in sea level in relation to land at approximately 30 cm per century. Since sea level has risen approximately 30 cm since the first dredging was completed in 1889, and dredging has increased high tide elevation by at least an additional 26.5 cm, the total rise in the vertical reach of tidal waters has increased 56.6 cm. Therefore, the potential extent of tidal inundation and seawater intrusion into the upper estuary over the last century has been significant.

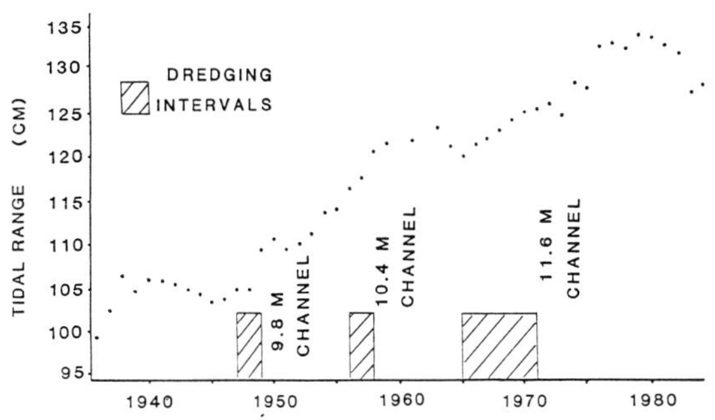

Figure 2. Tidal range at Wilmington, N.C. from 1936 to 1984 with dredging projects superimposed. Tidal data are from National Oceanographic Survey #2493 and dredging data from U.S. Army Corps of Engineers (1977).

Storm Surges

Hurricanes and tropical storms are an important environmental component in coastal North Carolina. Their frequency is irregular with a high frequency in the late 19th century (Neumann et al. 1978), with six hurricanes in the 1870's, five in the 1880's, four in the 1890's and four in the first ten years of the 20th century. Between 1910 and 1985 there were only seven hurricanes which passed near the Cape Fear region of North Carolina. Storm surges produced by tropical storms typically push flood waters into the upper Cape Fear estuary. Tropical storms, however, are usually accompanied by heavy rains and such storms typically decrease salinity in estuaries (Flemer et al. 1979; Turner et al. 1979). Extratropical storms, known locally as Northeasters, are more frequent, occurring more than 40 times a year (Boserman and Dolan 1968) and are not accompanied by heavy rains. Although less dramatic in terms of maximum tides they have a much greater opportunity to affect plant communities in the upper estuary because of their frequency and potential for pushing salt water up-river.

Wetland Communities in the Cape Fear Estuary

There are 15,000 acres of fresh and salt marsh and over 30,000 acres of tidal swamp along the Tidal portion of the Cape Fear River (Wilson 1962). Marshes along the lower portion of the estuary are primarily well-flooded salt marshes dominated by *Spartina alterniflora*. Brackish marshes dominated by *Juncus roemerianus* occupy large portions of the irregularly flooded marsh along the middle reaches of the estuary (Wilson 1962) (Fig. 1). Upper regions of the estuary are characterized by oligohaline (0-5 o/oo) marsh in abandoned

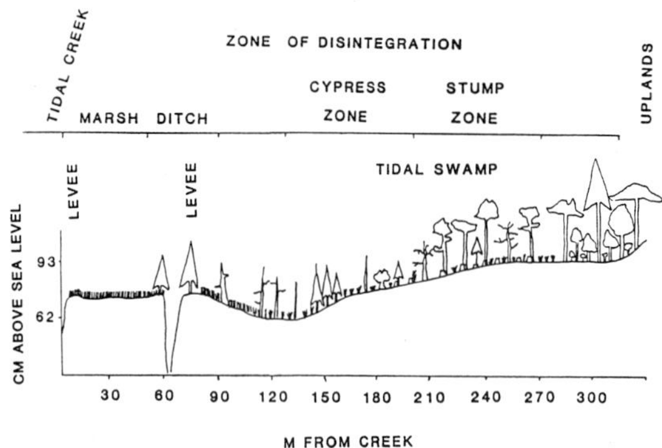

Figure 3. Composite of three transects through tidal swamp and tidal marsh in the upper estuary. Horizontal distance varied greatly, but the same zones were apparent in each transect. Vertical height is exaggerated and shows the height of these wetlands above sea level.

rice fields and by tidal swamp in undisturbed areas. Oligohaline marsh is dominated by Typha spp., Sagittaria lancifolia, Spartina cynosuroides, Zizania aquatica, Scirpus validus and Peltandra virginica with many other species present (Roza and Hackney 1984). Tidal swamps also contain a variety of species, but are dominated by bald bypress, (T. distichum), black gum (N. sylvatica), red maple (Acer rubrum) and water ash (Fraxinus caroliniana) (Rozas and Hackney 1984).

In upper regions of the estuary tidal swamps border both the river and abandoned rice fields. Vegetation in these swamps varies from uplands to tidal channels (Fig. 3). In wetlands adjoining upland communities, vegetation is thick and composed mostly of Cypress, Black Gum and Red Maple with an understory of Water Ash and saplings of the dominant species (Fig. 4A). This grades into a swamp composed of the same species without an understory. The majority of these trees are found growing on the top of old stumps and logs above the soil level. Few mature trees are found growing at the normal soil elevation. Shrubs such as Wax Myrtle (Myrica cerifera) and Swamp Rose (Rosa palustris) also growing on old stumps become more abundant toward the rivers and creeks. At the swamp edge there is typically a dense stand of young Cypress (< 20 years old) growing directly in the soil with many similar sized dead trees nearby in the marsh (Fig. 3 and 4B). This marks the beginning of the tidal marsh zone.

We interpret the disintegration of the tidal swamp and the growth of most healthy trees and shrubs in elevated locations (e.g. on old stumps) to be caused by increased tidal flooding and the presence of occasional inputs of more saline water. Nearby tidal marshes which are located in abandoned rice fields do not contain any woody species even though the elevation and soil of these communities are essentially the same. This suggests that old stumps from previously uncleared portions of tidal swamps are providing elevations above the tide line necessary for the growth of many of these freshwater species.

Figure 4A. Healthy tidal swamp in the upper estuary showing the dense nature of the vegetation which is mostly Red Maple (Acer rubrum) in this photograph.

Figure 4B. Border between tidal marsh and the Cypress zone of a tidal swamp in the upper estuary. Typha spp. in the foreground surround recently dead Black Gum (Nyssa sylvatica) and grade into a healthy stand of Bald Cypress (Taxodium distichum).

Figure 4C. Dead Cypress trees in a brackish marsh near Mott's Creek. Vegetation in the foreground is Typha sp. with the darker vegetation immediately behind Black Needlerush (Juncus roemerianus).

Estuarine and riverine wetlands typically maintain themselves during periods of rising sea level by accumulating sediments (Redfield 1972). Despite a 56.5 cm increase in the high tide level, swamps depicted in Fig. 3 are seldom flooded by more than 20 cm of water. Either these wetlands were well above the high tidal level in the past or they have been rapidly accumulating sediments. The presence of deep, unconsolidated soils containing mostly autochthonous, decaying organic matter (60-80% organic) suggests rapid accumulation.

Cypress stands have appeared along the ecotone between swamp and marsh (Fig. 3 and 4B). These relatively dense stands of trees are invariably associated with lower elevations (Fig. 3). Normal salinity of flooding water in this area is usually less than 1 o/oo, but may occasionally exceed 15 o/oo (Rozas and Hackney 1984). This is reflected by interstitial water salinity which is typically < 0.5 o/oo in healthy portions of the tidal swamp (Fig. 4A) and as high as 3.5 o/oo along the disintegrating areas of the tidal swamp and tidal marsh.

The presence of saline water indicates an increased input of soluble SO_4. Increased SO_4 in interstitial water has the potential for driving anaerobic sulfate reduction (Howarth and Hobbie 1982) which could increase decomposition of soils within tidal swamps. Lower elevations along tidal swamp borders where the Cypress zone is usually found (Fig. 3) may indicate that this is already occurring.

Aerial photography is an ideal way to document changes in wetland vegetation, especially when the change is dramatic and easily determined. Unfortunately, aerial photos prior to 1938 are few and of poor quality.

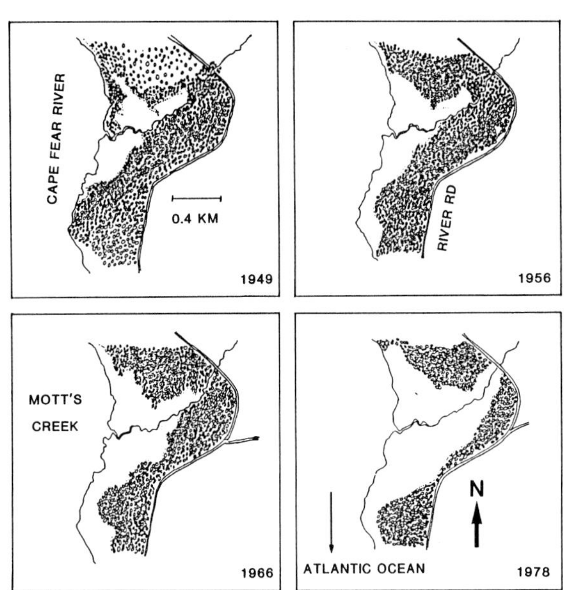

Figure 5. Mott's Creek in the middle reaches of the east side of the Cape Fear River (see Figure 1 for relative location). Based on Soil Conservation Service aerial photographs, they show a dramatic decrease in the extent of the forested areas. Most of the forests remaining west of the road in 1978 consist of upland pine forests. Most of the forested wetlands, which was probably tidal swamp, have been eliminated by 1978.

Dramatic changes in wetlands along the Cape Fear estuary are easily documented after 1938, especially for the middle reaches of the estuary. An area located on the east side of the river known as Mott's Creek is shown as an example (Fig. 5). Dead trees found today (Fig. 4C) are mostly Cypress with pines (Pinus spp.) along upland borders. Apparently, dramatic changes occurred (Fig. 5) when dredging of the river was increasing tidal amplitude (Fig. 2). Vast areas of tidal swamp were replaced by brackish marsh dominated by J. roemerianus and Scirpus olneyi (Fig. 5). Similar changes in wetland communities occurred throughout the middle reaches of the estuary.

Salt marshes dominated by S. alterniflora in the lower estuary appear to have been unchanged by the activities which have caused such dramatic changes upriver. Besides being close to the mouth of the estuary where tidal amplitude was probably already near that of the nearby coast, salt marshes are frequently able to maintain their position relative to sea level rise by accumulating sediments (Redfield 1972; McCaffrey and Thomson 1980).

DISCUSSION

It is clear from historical records that large changes have occurred in wetlands in the middle and upper reaches of the Cape Fear estuary during the past century. Parts of the river where salinity is now in the mesohaline range (5-18 o/oo) were once prime rice growing farmlands (Clifton 1973). Land which was flooded tidally by fresh water was considered the most valuable in the state (Clifton 1973). Today land which has these characterisitics exists far upriver of the estuary and shows no signs of having been used for rice farming. This again suggests that tidal flooding is relatively recent for these areas (after 1908).

In the oligohaline (0.5 - 5 o/oo) portion of the estuary vegetation changes in tidal swamps result from increased tidal flooding and salt intrusion. Tidal swamps are flooded by only 20 cm of water or less despite a 56 cm increase in high tidal level since 1885. Heavy accumulations of organic matter on the swamp floor and the growth of trees and shrubs on old stumps has allowed many plant species to persist in this altered ecosystem. Tidal swamps, however, seem to be in a state of transition. Elevated interstitial water salinities and frequent flooding during the growing season should stress many of the species currently inhabiting this community. Within the next 50 years many of these swamps now dominated by red maple and cypress will be replaced by tidal and brackish marshes dominated by various herbaceous species.

Changes observed in these wetlands, although not entirely natural, have been described as reverse succession in a freshwater wetland ecosystem in Minnesota (Isaak et al. 1959) and attributed to rising water levels. Although not described as such, a number of studies document reverse succession. Increased flooding causes a reversion to communities characteristic of early successional stages and in some cases even the loss of pioneer vascular plant species (DeLaune et al. 1978; DeLaune et al. 1983; Hatten et al. 1983; Baumann et al. 1984; Stevenson et al. 1985). In estuarine areas increased flooding is usually due to the inability of the plant community to produce enough organic material (autochthonous inputs) or trap enough sediments (allochthonous inputs) to increase or maintain marsh surfaces in the face of sediment compaction and sea level rise (Stevenson et al. 1986).

Changes within these tidal swamp communities is undoubtedly more complex than simply a change in species. Dramatic changes in sedimentation rates, decomposition rates of litter and seed germination are probably accompanying the rise in water level and salt concentration. Much work is needed before a full understanding of the relationship of these changes to the plant community and to the estuary as a whole can be reached.

REFERENCES

Boserman, K. and Dolan, R. 1968. The frequency and magnitude of extratropical storms along the outer banks of North Carolina. Technical Rpt. 68-4, Coastal Research Associates, Charlottesville, VA, USA. 72 p.

Baumann, R.H., Day, J.W. and Miller, C.A. 1984. Mississippi deltaic wetland survival: sedimentation versus coastal submergence. Science 224:1093-1095.

Clifton, J.M. 1973. Golden grains of white: Rice planting on the lower Cape Fear North Carolina Historical Review 50:356-393.

DeLaune, R.D., Patrick, W.H., Jr. and Buresh, J.B. 1978. Sedimentation rate determined by 137 Cs dating in a rapidly accreting salt marsh. Nature 27:532-533.

DeLaune, R.D., Baumann, R.H. and Gosselink, J.G. 1983. Relationship among vertical accretion, coastal submergence and erosion in a Louisiana Gulf marsh. Journal of Sedimentary Petrology 53:147-157.

Flemer, D.A., Ulanowicz, R.E. and Taylor, D.L. 1979. Some effects of tropical storm Agnes on water quality in the Patuxent River estuary. pp. 251-187. In The Effects of Tropical Storm Agnes on the Chesapeake Bay Estuarine System, CRC Pub. No. 54. Johns Hopkins University Press, Baltimore, MD, USA.

Hackney, C.T. and de la Cruz, A.A. 1978. Changes in interstitial water salinity of a Mississippi tidal marsh. Estuaries 1:185-188.

Hatten, R.S., DeLaune, R.D. and Patrick, W.H., Jr. 1983. Sedimentation, accretion and subsidence in marshes of Barataria Bay, Louisiana. Limnology and Oceanography 28:494-502.

Howarth, R.A. and Hobbie, J.E. 1982. Regulation of decomposition and microbial activity in salt marsh soils: A review. pp. 183-208 In Kennedy, V. S., ed. Estuarine Comparisons. Academic Press, New York, NY, USA.

Isaak, D., Marshall, W.H. and Buell, M.F. 1959. A record of reverse plant succession in a tamarack bog. Ecology 40:317-320.

McCaffrey, R.J. and Thomson, J. 1980. A record of accumulation of sediment and trace metals in a Connecticut salt marsh. Advances Geophysics 22:165-236.

Neumann, C.J., Cry, G.W., Caso, E.L. and Jarvinen, B.R. 1978. Tropical cyclones of the North Atlantic Ocean 1871-1977. U.S. Dept. of Commerce, National Oceanic and Atmospheric Administration, U.S. Govt. Printing Office, Washington, DC., USA. 170 p.

NOAA, 1985. Personal Communication. Data compiled by James R. Hubbard, U.S. Dept. of Commerce/NOAA, Rockville, MD, USA.

Oosting, H.J. 1956. The study of plant communities. W.H. Freeman & Co., San Francisco, CA, USA. 440 p.

Redfield, A.C. 1972. Development of a New England salt marsh. Ecological Monographs 42:201-237.

Rogers, G.C. 1970. The history of Georgetown County, South Carolina. University of South Carolina Press, Columbia, SC, USA. 332 p.

Rozas, L.P. and Hackney, C.T. 1984. Use of oligohaline marshes by fishes and macrofaunal crustaceans in North Carolina. Estuaries 7:213-224.

Stevenson, J.C., Kearney, M.S. and Pendleton, E.C. 1985. Sedimentation and erosion in a Chesapeake Bay brackish marsh system. Marine Geology 67:213-235.

Stevenson, J.C., Ward, L.G. and Kearney, M.S. (1986). Vertical accretion in marshes with varying rates of sea level rise. pp. 241-260. In Estuarine Variability, Wolf, E. D., ed. Academic Press, New York, NY, USA.

Turner, R.E., Woo, S.W. and Jitts, H.R. 1979. Phytoplankton production in a turbid temperate salt marsh estuary. Estuarine Coastal Marine Science 9:603-613.

U.S. Army Corps of Engineers, Wilmington District. 1977. Final Environmental Statement - Maintenance of Wilmington Harbor, North Carolina. U.S. Army Corps, Wilmington, NC, USA. 97 p. + 3 Appendices.

Waisel, Y. 1972. Biology of Halophytes. Academic Press, New York, NY, USA. 395 p.

Welcher, F.J. ed. 1963. Standard Methods of Chemical Analysis, 6th edition, V. 2, Part B. Van Nostrand Co., Inc., Princeton, NJ, USA. 2,541 p.

Wilson, K.A. 1962. North Carolina Wetlands: Their distribution and management. North Carolina Wildlife Resources Commission, Raleigh, NC, USA. 169 p.

THE RESPONSE OF ATLANTIC WHITE CEDAR WETLANDS TO VARYING LEVELS OF DISTURBANCE FROM SUBURBAN DEVELOPMENT IN THE NEW JERSEY PINELANDS

Joan G. Ehrenfeld

and

John P. Schneider[1]

Center for Coastal and Environmental Studies
Rutgers - The State University of New Jersey
New Brunswick, NJ 08903

ABSTRACT

The effects of suburban development on wetlands dominated by Atlantic white cedar were studied in the New Jersey Pinelands. Four groups of 4-5 sites each, representing different intensities of development impact, were sampled for hydrological regime, water quality, species composition and community structure. Hydrological regime was strongly affected by the proximity of dams, ditches or channels constructed in association with housing, but roads and housing had little effect alone. Levels of ammonia, ortho-phosphate, chloride, lead, and pH were elevated in developed sites compared with the very low values of all parameters in control sites. Species native to cedar swamps, particularly herbs, disappeared from impacted sites, while a diverse group of non-Pineland species, weeds, and exotics became established in these sites. Sites receiving direct road runoff were much more strongly affected than were sites near housing and/or roads but without direct stormwater inputs; however, the complexity of the gradient of suburban impact was reflected in the high variability among sites in each group. Target values for water quality parameters and species composition are given for use in assessing the quality of cedar wetlands subject to human disturbance.

[1]Current Address: U. S. Environmental Protection Agency, Chicago, IL.

INTRODUCTION

The incursion of suburban development into previously undisturbed areas brings with it many kinds of disruption to natural ecosystems. Among these are changes in the chemistry, the quantity and flow rates of ground and surface waters, and changes in the pool of species present in the region. The New Jersey Pinelands is a rapidly developing region, and it provides an excellent locale in which to examine the ecological effects of suburbanization. We have addressed our concern to the wetlands dominated by Atlantic white cedar (*Chamaecyparis thyoides* (L.) BSP), one of the major, highly-valued ecosystems of the Pinelands and one likely to be strongly affected by all of the various types of disturbance. We have therefore undertaken a study whose goals are: (1) to determine the relative importance of alterations in hydrology and water quality due to suburbanization in altering cedar swamps, (2) to determine the responsiveness of species composition and community structure to alterations in water quantity and quality, and (3) to determine whether different levels of suburbanization cause different levels of change in the cedar swamps.

The New Jersey Pinelands is an area of ca.4 x 105 ha that is congruent with the Miocene-age Cohansey Formation of the Outer Coastal Plain. Pitch pine (*Pinus rigida* Mill.) and oak (*Quercus alba* L., *Q velutina* Lam., *Q prinus* L.) forests dominate the uplands on the extremely sandy, acidic, nutrient-poor soils. (Forman 1979). Extensive lowland areas occupy topographic depressions and stream corridors, and support white cedar swamps, red maple (*Acer rubrum* L.) swamps, and pitch pine lowlands. The water table of the unconfined aquifer is shallow (often only 3 m and usually less than 15 m), so that infiltration is both rapid and responsible for 89% of surface water flow (Rhodehamel 1973). Infiltrating water remains at the top of the aquifer, so that 85% of this water surfaces in the wetlands that occupy every slight dip in the land surface. Because of the high permeability and low potential for nutrient

D. F. Whigham et al. (eds.), Wetland Ecology and Management: Case Studies, 63–78.

retention of the upland soils (Douglas and Walker 1979) and the strong hydrological connections between upland and lowland, pollutants and nutrients introduced into ground and surface waters from developments can be expected to travel with little attenuation to the swamps.

The effects of development on water resources have been frequently documented. Increases in nitrogen, phosphorus, cations, chlorides, heavy metals, and organic pollutants have been demonstrated for road runoff (Hunter et al. 1980; Gupta et al. 1981; Field 1985; Porcella and Sorenson 1980) as well as for septic tank leachate (Canter and Knox 1985; Groff and Obeda 1982; Brown et al. 1984; Sawhney and Starr 1977; Chen 1988; Katz et al. 1980). In the Pinelands, it has been found that leachate from septic tanks in upland areas is not accessible for uptake by native vegetation (Ehrenfeld 1987), nor is it removed or immobilized by the soil (Douglas and Walker 1979); it thus is potentially an important source of disturbance for nearby wetland communities. Table 1 shows some representative data comparing the baseline quality of Pinelands ground and surface water with the inputs of substances from urban runoff or septic field leachate. Changes in hydrology can be due to water withdrawals, the damming effect of road embankments, the building of drainage ditches or the channelization of the streams draining the wetland (Baxter 1977; Swales 1972; Maki et al. 1980).

The potential biotic responses to such changes in environmental conditions include virtually all aspects of community structure and function. In nutrient-poor systems, the increase in available nutrients due to pollution, together with the proximity of sources of seed from weeds, cultivars, and species usually confined to neighboring eutrophic sites may precipitate alterations in species composition. This has indeed been demonstrated for hardwood and mixed hardwood-conifer swamps (Ehrenfeld 1983), aquatic macrophytes (Morgan

Table 1. Typical nutrient concentrations in undisturbed Pinelands waters and in various sources of water pollution. Chemical quantities in mg l^{-1}.

Parameter	Ground Water*	Surface Water	Urban Runoff	Septic Leachate
T (^{0}C)	8.8 - 19.5	2.1 - 17.8		
pH	2.63 - 6.76	2.7 - 4.8	incr	7.3
Dissolved Solids	25 - 50	15 - 42	187 - 7,000	
BOD_5		0.2 - 0.9	1 - 700	143
Dissolved Organic C	2.5 - 12	6.0 - 17.0	1 - 150	
NO_3-N	0.001 - 0.15	0.00 - 0.07	0.01 - 1.5	0.2
NH_4-N	0.002 - 0.07	0.00 - 0.07	0.1 - 2.5	36.3
Organic N	0.03 - 0.50	0.07 - 0.44	0.01 - 16.0	7.3
Total P	0.003	0.00 - 0.02	2.39	15.1

*Source of data: Ground water: Durand and Zimmer 1982; Harriman and Voronin 1984; Surface water: Patrick et al. 1979, Urban Runoff: Porcella and Sorenson 1980, Septic leachate: Douglas and Walker 1979.

and Philipp 1986), and periphyton communities (Morgan 1987) of the Pinelands, as well as elsewhere (e.g., Haenck and Hengenveld 1981). Changes in community structure, involving increases or decreases in species abundance and dominance relations, and alterations in the proportionate representation of plant life-forms may occur. Similarly, aspects of community function such as net primary productivity, nutrient cycling patterns, or decomposition rates could reflect the net effects of disturbance on the growth of individual plants. Various classifications of the biological responses to development have been proposed (Odum 1985; Rapport et al. 1985), which all include the concept of effects on structural and functional properties of ecosystems; they also include the concept that increasing levels of stress from disturbance will cause different kinds of responses within a given ecosystem.

We have hypothesized that the response of cedar swamps to suburban development depends on the intensity of disturbance. Disturbance in areas subject to development can include, in order of increasing intensity of impact, the construction of unpaved roads in outlying portions of a development, the building of houses and septic systems on unsewered paved roads, and the diversion of road runoff through storm sewers directly into adjacent wetlands. Thus, we have used a series of sites to examine the responses of cedar swamps to this proposed gradient of disturbance from suburban development, in order to identify the most important aspects of hydrology and water quality that affect the swamps, and, conversely, to determine the most sensitive aspects of community structure and function that may be altered by this kind of human disturbance.

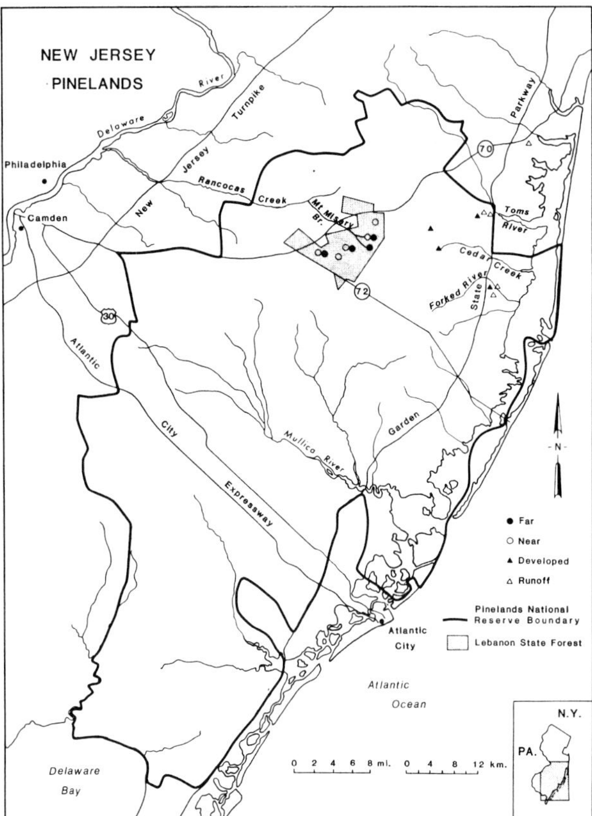

Figure 1. Map of southern New Jersey, showing the extent of the Pinelands and the locations of the study sites.

METHODS

Eighteen sites were selected within a 32 km diameter region in the northeastern section of the New Jersey Pinelands (Fig. 1). The sites met criteria of minimum density and size of the canopy trees (C. thyoides), geographical proximity, distance from the ocean, and proximity to source of disturbance. Four groups of sites (4-5 replicate sites per group) were identified: "C" or control sites were situated within Lebanon State Forest, at least 100 m from the nearest road; "N" or Near sites were also within the State Forest, but were adjacent to a road; "D" or developed sites were within housing developments that had paved roads, but neither sanitary nor stormwater sewers; and "R" sites were within housing developments that had paved roads with runoff piped through stormwater sewers directly into the swamp, but no sanitary sewers. Details of site selection and location are given in Schneider (1988) and Ehrenfeld and Schneider (1987). Additional background information on cedar swamps and the New Jersey Pinelands can be found in Laderman (1987), Forman (1979), Good and Good (1984), and Roman and Good (this volume).

Data on ground and surface water quality were obtained from monthly samples taken in each site. Ground water samples were taken from wells constructed of perforated PVC pipe and driven 30 cm into the peat substrate. Surface water samples were obtained when standing water was present in the hollows of the swamp. Dissolved oxygen was measured in situ with a YSI model 57 meter and a membrane-covered probe. Samples were analyzed for temperature and pH in the field, and returned to the laboratory for analysis of orthophosphate (o-PO_4) (ascorbic acid reduction, APHA 1981), ammonia (NH_3) (Solorzano 1969) and chloride (Cl) (mercuric nitrate method, APHA 1981). Heavy metals (Pb, Cu, Zn,) were determined on separate samples collected into acid-washed PVC bottles,

Table 2. Hydrological patterns of the four site types and sites with and without hydrological modifications. Means + standard errors, and coefficients of variation (C.V.) given for each parameter.

Site Type (n)	Annual Mean (cm)[1]		% Flooded[2]		# Floods		Range (cm)	
	Mean	C.V.	Mean	C.V.	Mean	C.V.	Mean	C.V
C (4)	1.85	4.38	74.8	0.09	3.5	0.37	38.0	0.29
	±0.49		±3.54		±0.65		±5.44	
N (5)	3.69	2.38	74.2	0.31	3.2	0.87	42.2	0.38
	±0.44		±10.23		±1.24		±7.14	
D (4)	0.73	7.35	79.8	0.10	3.0	0.27	26.5	0.92
	±0.30		±4.11		±0.41		±12.17	
R (4)	-0.18	78.91	79.3	0.16	±3.25	0.38	66.25	0.58
	±0.83		±6.41		±0.63		±19.27	
Unmodified	3.06	1.17	71.2	0.17	3.36	0.35	54.5	0.46
(11)	±1.08		±3.74		±0.36		±7.47	
Modified	4.38	0.86	87.2	0.11	2.50	0.42	22.5	0.58
(6)	±1.00		±4.17		±0.43		±5.33	

[1]over the 1982 sample year (12 months)
[2]over the 18-month study period

and analyzed by flame atomic absorption spectroscopy in the Environmental Sciences Laboratory, Rutgers University.

Water table fluctuations were measured using four 2-m long perforated PVC pipes emplaced in hollows at each site. Water table level was measured relative to the ground surface in the hollows weekly during April and May, biweekly during June-October, and monthly from November to March; measurements were made from April, 1982 until November, 1983.

The structure of the plant community was studied within a 20 x 30 m study plot at each site. Species richness was determined by monitoring the sites repeatedly during the 1982 growing season, and assigning each observed species a score on the Braun-Blanquet cover-abundance system (Mueller-Dombois and Ellenberg 1974). Twenty randomly-located 1 m^2 quadrats were used to quantify the abundance of herb, fern, and shrub stems, cedar seedlings, and the cover of Sphagnum spp. All trees (dead and live) over 2.5 cm DBH within a 1 m-wide strip around the quadrat were measured for DBH. Increment cores taken from ten dominant cedar trees per site (two cores per tree) were also obtained to measure tree growth rates.

Further details of chemical, hydrological and ecological methods are given in Schneider, (1988).

Data were analyzed using the Statistical Analysis System (Anon 1985) Parametric analyses of variance were used when preliminary testing of normality and homoscedasticity permitted; otherwise, nonparametric tests were used.

RESULTS AND DISCUSSION

Hydrological and Chemical Changes Along the Disturbance Gradient

We present below a summary of the specific changes in the physical and ecological parameters measured, as a basis for considering the general questions posed initially. Detailed presentations of the results are given in Ehrenfeld and Schneider(1987) and Schneider (1988).

The patterns of water table movement observed in the swamps are summarized in Table 2. In the control sites, the water table remained above the surface from early winter (December) until mid to late August, reaching a maximum of about 15 cm above the surface in April; it then fell to a minimum depth of 15-29 cm below the surface in October. Although the same general pattern was observed at the other sites, considerable variability within each class occurred, and could be attributed to alterations in drainage caused by the construction of dams, drainage ditches, and/or channelized streams. For the N and D site types, individual sites that had no structures present had a hydrological pattern not significantly different from that of the control group. In sites with downstream dams, the normal oscillation of the water table was highly damped; this is reflected in the smaller range of water table movement in the D sites (3 of 4 modified) and in the class of modified sites (Table 2), as well as in the reduced coefficient of variation of the annual mean for the modified sites. The water table stayed near the surface, fluctuating in some cases as little as 14 cm from maximum to minimum over the year.

In the R sites, more variable patterns were found. In sites without dams, the water table followed the normal yearly pattern, but with much larger ranges, reaching a depth of 55 cm and a height of over 20 cm. The larger fluctuations can be ascribed to the alternate flooding and draining effects of the storm sewer outfall and its associated dredged channel into the wetland. Like the modified sites in the other site groups, dammed sites had a damped range of values over the year, with the water table barely going below the surface. Thus, development affected wetland hydrology primarily through the placement in or near the wetland of structures that altered water flow. In the absence of specific structures, there was little impact that could be ascribed to the proximity of roads alone, the additions of water from septic field leachate, or the removal of water by homeowners' wells. The variable presence of structures modifying water flow within the site types contributed to the pattern of increasing variability in hydrological pattern seen in R sites (different types of modifications) compared with D sites (similar kinds of modifications) or undeveloped sites (Table 2; coefficients of variation).

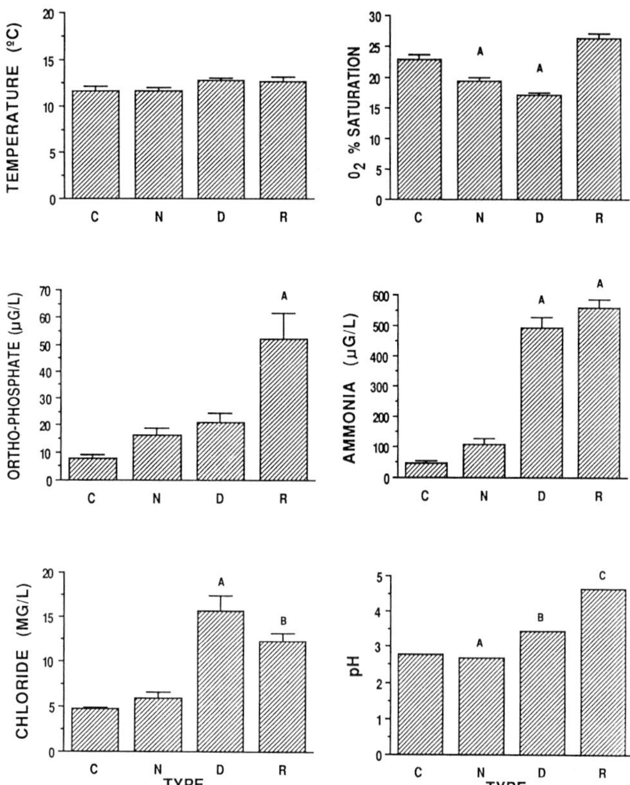

Figure 2. Ground water: annual mean values (and standard errors) for water quality parameters of the four site types. Means based on pooled samples from all sites within each group. Bars denoted by different letters are significantly different at $P < 0.05$

The annual means of the water chemistry variables in relation to level of development are shown in Figs. 2 and 3. Temperature was not significantly affected by any level of development. Dissolved oxygen in groundwater was decreased in N and D sites, reflecting decreases observed during the winter months. Surface water dissolved oxygen levels were decreased in D and R sites, reflecting patterns occuring during both summer and winter months (Larsen 1982). Decreases in the surface water oxygen concentrations may be due to decreased photosynthesis attributable to a loss of Sphagnum, an increase in uptake due to nutrient-induced microbial growth, or both. Chloride clearly identifies suburban water inputs to cedar swamps, in both groundwater and surface water. Differences in the patterns of occurrence of elevated chloride between D and R sites suggest, however, that the sediment/water interface within the swamp is an effective barrier to the movement of chloride. Thus, septic inputs are distinguishable from stormwater inputs.

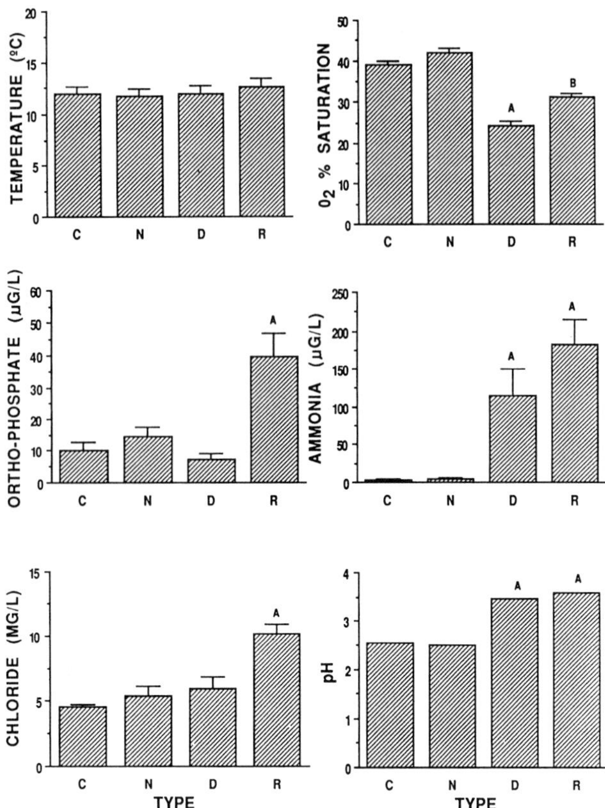

Figure 3. Surface water: annual mean values (and standard errors) for water quality parameters of the four site types. Means based on pooled samples from all sites within each group. Bars denoted by different letters are significantly different at $P < 0.05$.

Ortho-phosphate levels are at extremely low levels in undisturbed sites, and as low as or lower than those reported in classical bogs (Gorham et al. 1987; Schwinter and Tomberlin 1982). Only the R sites showed significant elevations of o-PO_4 in both ground and surface water. Ammonia-N concentrations, like chloride, reflect the presence of development. Extremely low values were observed in the surface waters of both control and N sites, while both surface and ground waters had elevated amounts (with up to 10-fold increases) in D and R sites. The increase in surface water concentrations in D sites suggests that unlike o-PO_4 or Cl, ammonia can move between ground and surface waters. Surface waters at all sites, however, have much lower concentrations than do ground waters.

Heavy metal concentrations in shallow ground water were determined in two samples; one followed a prolonged dry period, and the other followed rain (Table 3). Lead levels were significantly elevated at R sites during the dry period, but not after rainfall. No other significant differences were found, although the trend of the data for zinc levels also suggests an effect of added road runoff during dry periods.

Biological Changes Along the Disturbance Gradient

Species composition proved to be highly sensitive to development. All species observed in the study were classified by their "typical" habitat, using Stone's (1911) pre-development flora of the region as a standard. Using his descriptions, species were assigned to the following habitat categories: typical of cedar swamps; found in other Pinelands habitats; found in non-Pinelands habitats in New Jersey; exotic to the state. The results of this analysis are shown in Fig. 4. There is a clear trend towards increased species richness along the disturbance gradient. Increased species richness in wetlands has been associated with increases in water flow in other peatlands (Heinselman 1970), but in these sites, there was no uniform change in hydrology that matched the increase in species richness. Although the statistical analyses did not differentiate control and N sites, species from non-cedar swamp habitats occurred in N sites that never occurred in control sites. Thus, the data suggest that the presence of unpaved roads, with no other development, can introduce subtle changes into the swamps. Most of the added species present in the developed sites were herbs, and most were present at only one site and at low cover-abundance values (Schneider and Ehrenfeld 1987). Vine species, though low in number, were a conspicuous addition to the flora of the D and R sites.

Structural aspects of the cedar swamp communities did not vary along the development gradient. Cedar tree densities (in the control sites, mean of 5,676 ± 1,040 stems > 2.5 cm DBH ha^{-1}) were high, as is typical for this species (Forman 1979). Several of the D and R sites had, however, higher numbers of cut stumps and lower numbers of large trees than control sites, suggesting that illegal removal of large trees may be an impact associated with suburbanization. Shrub stems (229,000/ha ± 39,000) also are very dense, as they are in other forested wetlands of the Pinelands (Ehrenfeld and Gulick 1981; Ehrenfeld 1986). Herbaceous plants ($8.4/m^2 \pm 2.9$) are patchily distributed, giving rise to high variability within each site type.

Two components of community structure

Table 3. Heavy metal concentrations ($\mu g\ l^{-1}$) in groundwater following wet and dry conditions. Statistical differences shown by significance level of the ANOVA; letters indicate significantly different means using the SNK test. Sample size in parenthesis.

Metal	P	Control (4)	Near (4)	Developed (4)	Runoff (4)
Lead					
Wet	ns	1.3 ± 0.4	1.8 ± 0.4	1.1 ± 0.4	6.9 ± 2.9
Dry	.005	2.5 ± 0.8	3.7 ± 1.7	15.8 ± 4.5	119.8 ± 36.0^{A}
Zinc					
Wet	ns	33.8 ± 8.0	32.5 ± 7.2	19.5 ± 6.4	22.3 ± 5.4
Dry	ns	70.0 ± 18.9	61.3 ± 19.8	71.3 ± 21.1	267.5 ± 105.9
Copper					
Wet	ns	4.8 ± 1.1	4.0 ± 0.6	3.5 ± 0.9	3.2 ± 0.4
Dry	ns	2.6 ± 2.0	7.8 ± 3.9	7.3 ± 3.6	0.9 ± 0.3

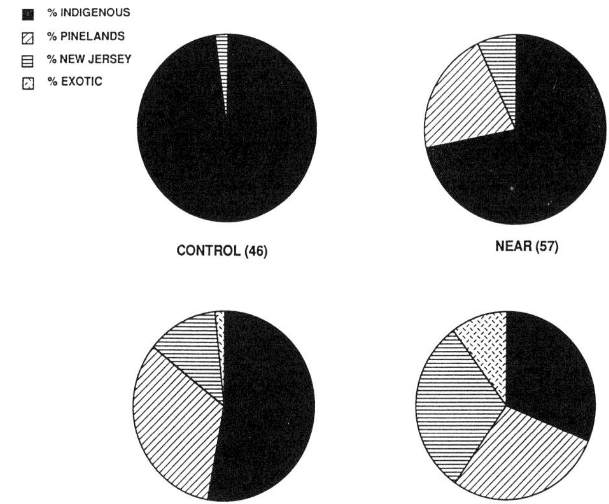

Figure 4. Percentage representation of species from different habitat types within each site type. Numbers in parentheses are the total species richness within each site type; *indicates significant difference at P < 0.05.

did respond to the disturbance from development. One was the amount of cover by Sphagnum spp., which fell from a mean of 81 ± 6% in control sites to 10 ± 4% in runoff sites ($P < 0.05$). N and D sites, while having cover values intermediate between these extremes, were not separated by the means test. The other component was the density of cedar seedlings, which was 20.3 ± 6.8/m^2 in control sites, and 2.4 ± 0.6/m^2 in runoff sites ($P < 0.05$); N and D sites again were intermediate in value.

Growth rates of dominant cedars (mean annual increment over the past 10 yr), which is an indicator of net primary production, were not significantly different in developed versus control sites. However, the mean ages of the trees in the different site types were different (84.6 ± 2.5 yr in control, versus 53.0 ± 2.0 yr in R sites), and this could have masked changes in age-specific growth rates. Therefore, mean annual increment over the past 20 yr was analyzed for the subset of trees aged 50-70 yr within each site; this analysis showed a significant ($P < 0.05$) increase in growth rate for trees in R sites compared to trees in all other site types.

Table 4. Correlations between principal component scores and original variables. All correlations at P< .001 or less.

Dataset	Axis 1	r	Axis 2	r	Axis 3	r
Hydrology	Var. Level (all)	.93	Mean level (all)	.81	Inflections	.73
	Range, level (all)	.93	Mean, Gr-82	.77	Flooded, winter	.75
	Range, Gr-82[1]	.88	Mean, Gr-83	.82	Flooded, winter	.63
	Var., Gr-82	.87				
	Range, Gr-83	.87				
	Var., Gr-83	.79				
	Range, winter	.90				
	Var. winter	.87				
Water Chemistry	Mean gw NH_3[2]	.82	Mean, sw CL	-.75	Mean gw T^oC	-.50
	Range sw NH_3	.84	Var, sw CL	-.75	Range, gw DO	.59
	Var, gw NH_3	.81	Mean, gw Cl	-.65	Var. gw T^0C	.54
	Mean, gw H[3]	-.79	Range, sw Cl	-.71	Wet Cu	.50
	Mean, gw PO_4	.77	Range, sw Cl	-.71		
	Range, gw H[3]	-.79	Wet Pb	-.61		
Species Affinity	Total invaders	.99	Pine Barrens spp.	.82		
	Relative diversity	-.96	Indigenous spp.	.81		
	New Jersey spp.	.92	Cedar Swamp spp.	.44		

[1] Gr = growing season; 82 = 1982 data, 83 = 1983 data

[2] gw = ground water; sw = surface water

[3] analyses were performed using hydrogen ion concentrations

Relationships among environmental and biotic variables

Principal components analysis was used to identify those variables or sets of variables which could best be used to signal the presence of disturbance from development, through correlations of the axis scores with the original variables. These analyses were also used to examine the validity of the hypothesized gradient of disturbance, through examination of the distributions of the sites within the ordination spaces.

Principal components analyses were carried out using SAS programs on standardized data. Analyses of hydrological data used seasonal (growing season, April-September, and winter, October-March) as well as annual means, variances, and ranges. Analyses of the water quality data used only annual means, because of missing data due to the absence of surface water in some sites in some months, and the loss of some samples.

Fig. 5 shows the ordination of the sites based on the water chemistry data. Three axes accounted for a total of 63% of the variation in the data. Table 4 gives the correlations of the variables with the principal component scores. The first axis (29.3% of the variance) separated sites within developments (D and R sites) from sites within undeveloped watersheds (C and N sites), along gradients of ammonia, pH and groundwater phosphate. The analysis also showed that the amount of variation in these parameters was important in separating the sites; range and variance of the key parameters were as highly correlated with axis scores as were annual means. The second axis, (19.9% of the variance) also reflected the disturbance gradient, separating the sites with high chloride

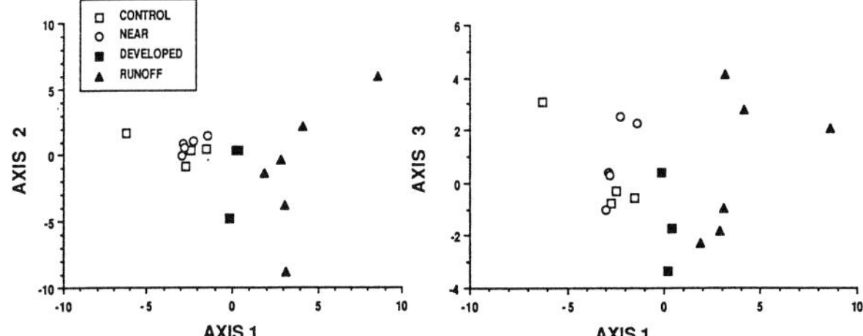

Figure 5. Principal components ordination of sites based on water quality variables included in each axis are given in table 4.

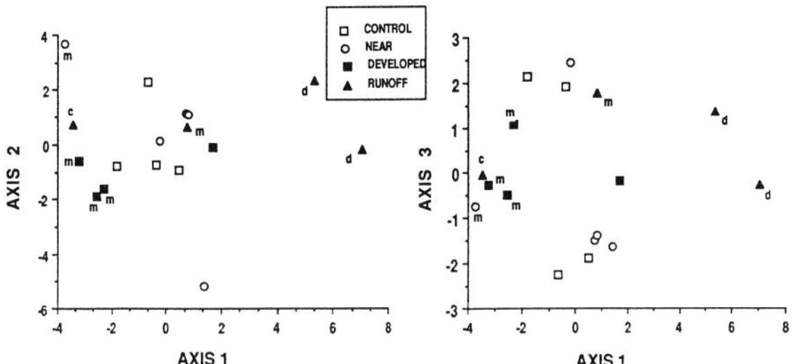

Figure 6. Principal components ordination of sites based on hydrological variables. m = sites with dams; c = sites with channelized streams; d = sites with drainage ditches. Variables included on each axis are given in table 4.

and high lead (outliers at opposite ends of this axis) from the undisturbed sites clustered in the middle. Again, variance and range of the variables were as important as the mean values. Surprisingly, the third axis (13.7% of the variance) was correlated with temperature, despite the fact that analysis of the site means revealed no changes along the gradient. However, this third axis was sensitive to variance in temperature, as well as the mean, and the sites were distributed as on the other axes, with C and N sites clustered together in the middle, while D and R sites were near the ends of the axis.

Ordination of the sites using the hydrological variables is shown in Fig. 6. The first three axes accounted for 82% of the variance. However, none of the axes effectively separated sites within developments (D and R sites) from the control and N sites. Rather, the ordination illustrated several patterns of swamp hydrology unconnected to the proximity of housing. The first axis (47% of the variance) reflected the range of water table depths, rather than the mean level. Sites with dams, as described above, were grouped at one end of the axis, while unmodified sites from

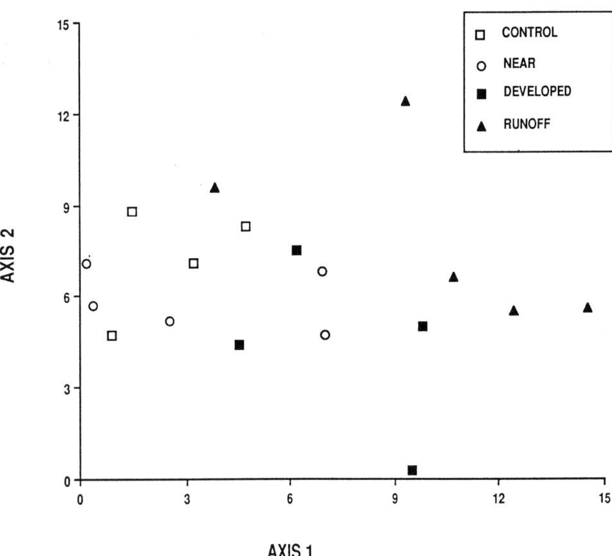

Figure 7. Ordination of sites using DECORANA, based on species presence within each site.

undisturbed watersheds were clustered in the center. The two runoff sites with dredged channels, and extreme ranges of water table change, occupied the other end of the axis. However, the second axis (22% of the variance) reflected strong correlations with the mean water table level. The distribution of sites along this axis suggested that natural variations in the wetness of sites is independent of the intensity or presence of development. Lowry (1984) similarly found considerable variability in the water table level of undisturbed cedar stands. The third axis (13% of the variance) was corrrelated with the number of flooded periods during the study; again, there was no relationship to the disturbance gradient.

Thus, the ordination analysis of the hydrological data indicates that there is considerable variation in the water table dynamics of cedar swamps independent of the effects of suburban development, as well as variability attributable to the construction of structures that modify water flow. Such structures alter the variability in water table depth and affect the extremes of flooding and drying over the annual cycle, rather than affect the mean water table level. However, the wetness of a swamp, as reflected in the mean water table depth and the tendency for fluctuations to include alternations of flooded and unflooded periods, varies independently of human impact.

Analysis of the biotic variation among sites was restricted to the data on species

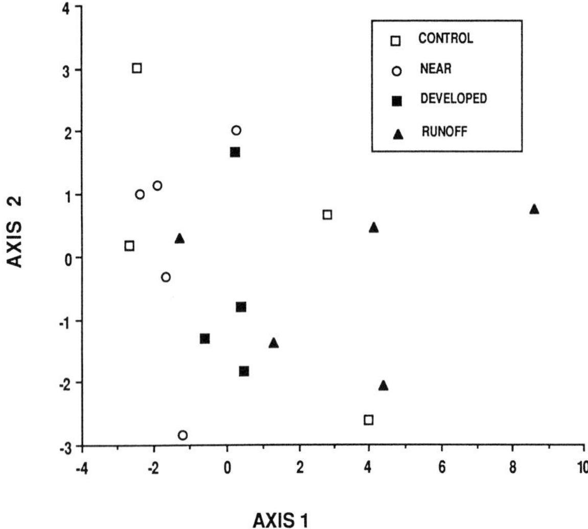

Figure 8. Principal components ordination of sites based on biogeographic/habitat indices of species composition within each site. Variables included on each axis are given on table 4.

composition, because so few differences in structural variables were observed among the sites in the initial analyses. The data on species composition were ordinated in two ways: (1) the complete species x site matrix was submitted to detrended correspondence analysis (DECORANA; Gauch 1982), and (2) a variety of indices reflecting the biogeographic composition of the flora at each site was ordinated using the principal components analysis program in SAS. These indices are fully described in Ehrenfeld and Schneider (1987).

Figure 7, the analysis of the species composition by DECORANA, shows that there is a clear clustering of the C and N sites together at one end of the first axis. The D and R sites are, however, more broadly spread out over the ordination space. D sites form a bridge between the cluster of sites from undeveloped areas and the widely separated points of the R sites. There is sufficiently little variation in ordination scores along the second axis to warrant interpretation. Thus, this analysis suggests that while proximity to roads has little effect on cedar wetland species composition, proximity to suburban developments has a strong effect, and the impact increases with the presence of direct stormwater inputs.

Ordination of the sites based on the biogeographic affinities of the species at each site gave the results shown in Fig. 8. The first two axes accounted for 75% of the variance. The first axis (57%) separated the sites in a manner similar to that of the ordination based on species lists: control and N sites are clustered together at one end of the gradient, and D sites form a bridge connecting this cluster and the more widely dispersed points representing the R sites. This axis was correlated with variables that expressed the presence of species not normally found in cedar swamps, and the proportion of the total species richness at a site attributable to native swamp species; i.e., it expressed the incursion of undesirable species into the wetlands. It is noteworthy, however, that the presence of suburban development does not necessarily cause such changes, because several of the D sites were located adjacent to N sites in the ordination space. The second axis (18% of the variance), separated sites according to the number of native species present. Rather than separate sites along the disturbance gradient, this axis apparently reflects one aspect of natural variation in species diversity among the population of sites. Some undisturbed sites have high species richness, while other equally undisturbed sites have a low diversity of native species. Differences in site history, variations in micro-environment, differences in the nature of the surrounding communities, differences in wetland size, as well as random factors associated with plant recruitment and establishment are possible explanations for this variability.

Perusal of the ordinations (Figs. 5-8) shows that on almost all axes of all the analyses, the disturbed sites were more widely spread through the ordination space than were the undisturbed sites. This observation was tested by calculating the coefficient of variation of the axis scores for two groups of sites (C and N versus D and R). The sites were lumped into two, rather than four groups, because none of the ordinations suggested a strong separation between C and N or between D and R. Table 5 demonstrates that in all but two cases (Hydrology Axis 2 and Species Affinity Axis 2), the coefficients of variation of the developed sites (D and R) are much higher than those of the undeveloped sites. Thus, one of the salient features of the developed sites is a greater variability in the distinguishing features of both biotic and physical characteristics. Suburban development thus not only alters the absolute values of environmental and biological attributes of these wetlands, but also decreases the degree of similarity among swamps.

These data suggest, furthermore, that there may be a causal relationship between the increase in variability of environmental characteristics and the changes in species

Table 5. Mean (± standard error) and coefficients of variations (C.V.) of principal component scores for disturbed (D and R) and undisturbed (C and N) sites.

Dataset	Axis	Undisturbed Sites			Disturbed Sites		
		Mean	Standard Error	C.V.	Mean	Standard Error	C.V.
Hydrology	1	-0.369	±0.526	4.282	0.415	±1.425	9.714
	2	0.059	±0.788	39.964	-0.078	±0.482	15.678
	3	-0.326	±0.637	5.859	0.031	±0.011	30.70
Water Chemistry	1	-2.80	±0.471	0.504	2.801	±0.850	0.910
	2	0.610	±1.271	0.258	-0.607	±1.428	6.657
	3	0.659	±0.517	2.350	-0.657	±0.993	4.543
Species Affinity	1	-2.063	±0.394	0.574	1.809	±0.394	1.672
	2	0.248	±0.650	7.857	-0.223	±0.514	6.461

composition discussed above. Suburban development evidently influences the physical environment of different sites in different ways, such that the environmental characteristics of previously similar sites diverge. Thus, some may undergo increases in one set of chemical parameters while other sites experience changes in a different constellation of chemical variables. A similarly varying range of effects may occur in water table relationships such that sites within developments may continue to resemble undisturbed swamps, or may change in one of several different ways. Thus, the particular set of chemical and hydrological conditions produced at a given site as suburban development occurs may create habitat for a small subset of the total list of species recorded at developed sites. The changes in species composition support this interpretation: each disturbed swamp has a different group of weeds, exotics, and other wetland plant species present. In this way, variability in the effects of development among sites may foster the variability in species composition seen in this dataset.

CONCLUSIONS AND MANAGEMENT RECOMMENDATIONS

The Nature of the Disturbance Gradient

The initial hypothesis used to group the sites was that there is an increasing amount of disturbance from development that accompanies, in turn, the creation of roads, the building of unsewered houses, and the introduction of runoff from paved areas directly into wetlands. Significant differences among site types for the environmental and biotic variables reported above all suggest that the runoff (R) sites are much more severely affected than are the D or N sites. No statistically significant differences were found between the control and the N sites, although small suggestive differences in species richness and the presence of non-native species were noted. The lack of statistically significant differences between the undeveloped (C and N) sites and the D sites probably reflects the great amount of variation in site characteristics within the D group displayed in the ordination studies. The analyses furthermore suggested that variation among sites itself may be at least as biologically important as the deviations in the absolute values of the variables, as discussed above.

A stepwise discriminant analysis was applied to the data, in order to determine whether a simple set of variables could be used to assign sites to categories of development impact, as hypothesized. The analysis was conducted using factors extracted from each dataset by factor analyses. These factors were used in order to reduce the number of variables in each analysis. In order to distinguish the four site groups with a minimum probability of 0.97, the discriminant function analysis required seven factors, involving a wide range of chemical, hydrological, and biotic variables. Thus, the gradient of disturbance is very complex, such that assignment of a site to a category of disturbance level requires information about chemical, hydrological, and biological features of the site.

In summary, while white cedar swamps respond to different intensities of development in variable and complex ways, the effects of direct road runoff inputs into the wetland stand out as much more severe than other modes of suburban construction. Sites within developments are likely to experience altered hydrological regimes due to the building of structures or ditches, they are likely to have some changes in water quality, and they are likely to have well-defined changes in the species composition of the herb layer. Sites additionally exposed to road runoff have much more marked alterations of water chemistry and species composition than do other developed sites. Sites in undeveloped watersheds, but proximate to roads, may experience slight changes in species composition but few effects on water chemistry or flow.

Table 6. Range of mean annual water chemistry values from C sites, to be used as a standard of comparison.

	Parameter	Range		
Groundwater				
	pH[1], mean	2.53	-	2.92
	pH, variance	4.97	-	6.55
	pH, range	1.96	-	2.73
	Cl, mean (mg l^{-1})	3.57	-	5.70
	Cl, variance	0.55	-	3.42
	Cl, range	2.83	-	7.27
	NH_3, mean (µg l^{-1})	6.82	-	80.72
	NH_3, variance	441.3	-	12886.0
	NH_3, range	70.7	-	342.1
	PO_4, mean (µg l^{-1})	3.31	-	11.58
	PO_4, variance	135.0	-	466.69
	PO_4, range	49.7	-	91.6
Surface Water				
	pH[1], mean	2.28	-	2.72
	pH,[1], variance	4.21	-	6.01
	pH, range	1.57	-	2.46
	Cl, mean (mg l^{-1})	3.42	-	5.21
	Cl, variance	0.64	-	5.93
	Cl, range	2.98	-	9.11
	NH_3, mean (µg l^{-1})	0.00	-	9.05
	NH_3, variance	0.00	-	612.0
	NH_3, range	0.00	-	73.3
	PO_4, mean (µg l^{-1})	3.56	-	18.16
	PO_4, variance	104.1	-	1584.3
	PO_4, range	20.0	-	119.2

[1]All pH parameters calculated as (H+) and reconverted to pH.

Relative importance of chemical and physical variables

Both the univariate analyses of water chemistry and the multivariate analyses identified pH, ammonia, phosphate, chloride and lead, in ground and surface water, as important factors separating disturbed from undisturbed sites. Both winter and summer values were important, and both contributed to differences in annual means reported here. The principal components analysis also suggested that the variation in these parameters was important in describing the amount of impact from development.

Hydrological changes were tied to physical modifications of the environment associated with construction; the types of changes were thus dependent on the kind of hydrological disturbance present, rather than on the proximity of development itself. Sites within developments were clearly much more likely to have such physical alterations of the environment present than sites in undeveloped watersheds. Changes in the range of water table fluctuation, more than changes in mean level, can serve as indicators of the presence of flow-modifying structures. Thus, wetland hydrology shows little direct sensitivity to the presence of housing, but is responsive to the associated secondary likelihood that dams, ditches, and diversions will be constructed near the housing. The nature of the effect of altered hydrology on the wetland will also not reflect the size or proximity of the development, but rather will be specific to the kind of structure that is built in the wetland.

From these considerations, it is apparent that water chemistry provides a more useful group of variables that can be used to monitor or assess the degree of impact on particular wetland sites. The similarity of the patterns of site distribution within the ordination analyses of water chemistry, species composition, and biogeographic affinity also suggests that water chemistry variables would be more useful in assessing potential risk to plant community composition than would the hydrological variables. We suggest that the range of values for each water quality parameter observed in the control sites be used as a criterion against which to evaluate the quality of cedar wetland sites affected by suburban development; these values are listed in Table 6. Although there was little indication in the data of a distinction between the N and control sites, the latter are known to be free of suburban influence, and thus provide the clearest standard against which to judge other areas. The finding that variability in chemical parameters is as important as the mean values in distinguishing sites indicates that sampling programs should include repeated measurements, taken throughout the year.

Relative importance of biotic variables

The various analyses of biotic structure demonstrate that the influx of species not native to the swamps can be used as a sensitive indicator of the degree of disturbance already present, and that structural aspects of the plant community are mostly insensitive to adjacent development. We suggest that three measures can be used to characterize the quality of other sites, based on a complete list of vascular plant species compiled for the site. The list should be based on surveys made throughout the growing season, in order to adequately census early and late-blooming species. First, from this list, the number of species that, in Ehrenfeld and Schneider (in press), are classified as N, ND or NDR, should be tabulated; this number represents species found in sites with at least some impact from development. Among the sites studied here, no more than 12% of the species at any "near" site belonged to this category. We suggest, therefore, that a site with over 15% representation of species from this category be considered to reflect some disturbance. Second, the number of species on the site list should be again compared with Ehrenfeld and Schneider (in press), and the number of species categorized as DR, D or R be determined. This group represents species found only in sites within developments; the presence of any species from this group within a wetland should be considered evidence of substantial disturbance. As the percentage representation of the species in this latter group increases, the impact of disturbance should be considered increasingly great. Third, the biogeographical information in the lists of Ehrenfeld and Schneider (in press) should be used to calculate the total number of "invader" species (those not native to cedar swamps specifically), the total number of "indigenous" species (those found typically in cedar swamps), and the ratio of the difference between indigenous and invader species to the total species richness. These values can be compared with reference values given in Table 7 to evaluate the amount of impact evident at a given site.

In conclusion, suburban development affects water chemistry, water flow and species composition of cedar wetlands. There is a general trend for increasing impact with increasing intensity of development, but the relationship is not linear, and does not hold for every variable measured in this study.

Table 7. Ranges of values for biodiversity indices in C and N sites that can be used as standards for comparison.

Index	Control Sites	Near Sites
Indigenous Species[1]	17 - 36	16 - 29
Invader Species[2]	0 - 1	1 - 9
Relative native value[3]	0.92 - 1.00	0.53 - 0.93

[1]Defined as species occurring only in C sites and also identified by Stone (1911) as being typical of cedar swamps.

[2]Defined as species occurring in other Pinelands habitats, other New Jersey habitats, or exotic to the state.

[3] Defined as the ratio: (Indigenous-Invader)/(Indigenous + Invader).

Sites directly receiving stormwater runoff are much more strongly altered than are other developed sites; sites affected solely by roads are only distinguishable from undisturbed sites by the occasional presence of a few species not normally found in cedar swamps. Variability in environmental parameters may be as important as changes in the absolute values of these parameters. Species composition is the most sensitive indicator of disturbance; plant community structure is relatively insensitive to development impacts. Changes in water chemistry and hydrology can be pronounced, but with less capability of characterizing the level of impact, in part because of high variability in the actual impacts among sites. Naturally occurring variation among sites, especially with regard to hydrology, complicates the interpretation of site assessments based purely on physical characteristics. Thus, the presence of disturbance in cedar wetlands and the assessment of its intensity can be ascertained through detailed surveys of its species composition and water chemistry, and, to a lesser extent, knowledge of water-diverting structures in the drainage basin of the wetland.

ACKNOWLEDGEMENTS

We thank David Snyder, Shai Reisfeld, and Rebecca Mulcahy for assistance with the field and laboratory studies. Financial support came from the N. J. Department of Environmental Protection, the U. S. Department of the Interior, U. S. Geological Survey, and the Division of Pinelands Research, Center for Coastal and Environmental Studies, Rutgers University.

REFERENCES

American Public Health Association. 1981. Standard methods for the examination of water and wastewater, 15th ed. American Public Health Assocation, Washington,D.C

Anon.1985.SAS User's guide: Basics and Statistics, Version 5, edition, SAS Institute, Cary, NC. USA. 1290 p.

Baxter, R. M. 1977. Environmental effects of dams and impoundments. Annual Review of Ecology and Systematics 8:255-83.

Brown, K. W., Donnelly, K. C., Thomas, J. C., and Slowey, J. F. 1984. The movement of nitrogen species through three soils below septic fields. Journal of Environmental Quality 13:460-465.

Canter, L. W. and Knox, R. C. 1985. Septic Tank System Effects on Ground Water Quality. Lewis Publishers, Inc., Chelsea, MI. USA. 500p.

Chen, M. 1988. Pollution of ground water by nutrients and fecal coliforms from lakeshore septic tank systems. Water, Air and Soil Pollution 37:407-417.

Douglas, L. A. and Walker, R. 1979. Non-point

sources of pollution and soils of the Pine Barrens of New Jersey. New Jersey Agricultural Experiment Station. New Brunswick, NJ. USA. 81 p.

Durand, J. B. and Zimmer, B. J. 1982. Pinelands surface water quality, Part I. Center for Coastal and Environmental Studies, Rutgers University, New Brunswick, NJ. USA. 196p.

Ehrenfeld, J. G. 1983. The effects of changes in land-use on swamps of the New Jersey Pine Barrens. Biological Conservation 25:353-375.

Ehrenfeld, J. G. 1986. Wetlands of the New Jersey Pine Barrens: the role of species composition in community function. American Midland Naturalist 115:301-313.

Ehrenfeld, J. G. 1987. The role of woody vegetation in preventing groundwater pollution by septic tank leachate. Water Research 21:605-614.

Ehrenfeld, J. G. and Gulick, M. 1981. Structure and dynamics of hardwood swamps in the New Jersey Pine Barrens:contrasting patterns in trees and shrubs. American Journal of Botany 68:471-481.

Ehrenfeld, J. G. and Schneider, J. P. in press. Chamaecyparis thyoides wetlands and suburbanization: effects on hydrology, water quality and plant community composition. Journal of Applied Ecology.

Field, R. 1985. Urban runoff: pollution sources, control and treatment. Water Resource Bulletin 21:197-206.

Forman, R. T. T. 1979. Pine Barrens: Ecosystem and Landscape. Academic Press, NY. USA. 601p.

Gauch, H. G., Jr. 1982. Multivariate analysis in community ecology. Cambridge University Press, New York, NY. USA. 298 p.

Good, R. and Good, N. F. 1984. The Pinelands National Reserve: an ecosystem approach to management. BioScience 34:169-176.

Gorham, E., Eisenriech, S.J., Ford, J. 1985. The Chemistry of Bog Waters. pp. 339-359. In Stumm, W., ed. Chemical Processes in Lakes. John Wiley, New York, NY. USA.

Groff, D. W. and Obeda, B. A. 1982. Septic performance in hydrologically sensitive soils. Wetlands 2:286-302.

Gupta, M. K., Agnew, R. W., and Kobriger, N. P. 1981. Constituents of Highway Runoff. Vol. I: State of the Art Report. U.S. Federal Highway Administration, Report No. FHWA/RD-81/042. Washington D.C. USA. 111p.

Haeck, J. and Hengeveld, R. 1981. Changes in the occurrence of Dutch plant species in relation to geographical range. Biological Conservation 19:189-197.

Harriman, D. A. and Voronin, L. M. 1984. Water Quality Data for Aquifers in East-central New Jersey, 1981-1982. U.S. Geological Survey Open-file Report 84-281, Trenton, NJ. USA. 39p.

Heinselman, M. L. 1970. Landscape evaluation, peatland types, and the environment in the Lake Agassiz Peatland Natural Area, Minnesota. Ecological Monographs 40:235-261.

Hunter, J. V., Balmat, J., Wilber, W., and Sabatino, T. 1980. Hydrocarbons and heavy metals in urban runoff. pp. 22-41. In Jones, R.C., Redfield, G.W. and Kelso, D.P., eds. George Mason University. Urbanization, Stormwater Runoff and the Aquatic Environment. Fairfax, VA. USA.

Katz, B.G., Linder, J.B. and Ragone, S.E. 1980. A comparison of nitrogen in shallow ground water from sewered and unsewered areas, Nassau County, New York, from 1952 through 1976. Ground Water 18:607-616.

Laderman, A.D. 1987. Atlantic White Cedar Wetlands. Westview Press, Boulder, CO. USA. 401 p.

Larsen, J.A. 1982. Ecology of the Northern Lowland Bogs and Conifer Forests. Academic Press,New York, NY.USA. 307p.

Lowry, D.J. 1984. Water regimes and vegetation of Rhode Island forested wetlands. Master's Thesis, University of Rhode Island. Kingston, RI. USA. 174 p.

Maki, T. E., Weber, A. J., Hazel, D. W., Hunter, S. C., Hyberg, B. T., Flinchum, D. M.,Lollis, J. M. Rognstad, J. B., and Gregory, J. D. 1980. Effects of stream channelization on bottomland and swamp forest ecosystems. Water Resources Research Institute, University of North Carolina. Report No.147. Raleigh, NC. USA. 135p.

Morgan, M. D. 1987. Impact of nutrient enrichment and alkalinization on periphyton communities in the New Jersey Pine Barrens. Hydrobiologia 144:233-241.

Morgan, M. D. and Philipp, K. D. 1986. The effect of agricultural and residential development on aquatic macrophytes in the New Jersey Pine Barrens. Biological Conservation 35:143-158.

Mueller-Dombois, D. and Ellenberg, H. 1974. Aims and Methods of Vegetation Ecology. J. Wiley and Sons, New York. NY. USA. 547p.

Odum, E. P. 1985. Trends expected in stressed ecosystems. BioScience 35:419-422.

Patrick, R., Matson, B., and Anderson, L. 1979. Streams and lakes in the Pine Barrens. pp. 169-194. In Pine Barrens:Ecosystem and Landscape. Forman, R.T.T., ed. Academic Press, New York. NY. USA.

Porcella, D. B., and Sorensen, D. L. 1980. Characteristics of Nonpoint Source Urban Runoff and its Effect on Stream Ecosystems. EPA 600/3-80-032, Corvallis, OR. USA. 99p.

Rapport, D. J., Regier, H.A. and Hutchinson, T. C. 1985. Ecosystem behavior under stress. American Naturalist 125:617-640.

Rhodehamel, E. C. 1973. Geology and water resources of the Wharton Tract and the Mullica River Basin in southern New Jersey. New Jersey Division of Water Resources. Special Report. 36. Trenton, NJ. USA. 58p.

Sawhney, B.L. and Starr, J.L. 1977. Movement of phosphorus from a septic system drainfield. Journal of Water Pollution Control Federation. 49:2238-2242.

Schneider, J.P. 1988. The Effects of Suburban Development on the Hydrology, Water Quality and Community Structure of Chamaecyparis thyoides (L.) B.S.P. Wetlands in the New Jersey Pinelands. Ph.D. Dissertation, Rutgers University. New Brunswick, NJ. USA. 140 p.

Schneider, J.P. and Ehrenfeld, J.G. 1987. Suburban development and cedar swamps: effects on water quality, water quantity and plant community composition. pp. 271-288. In Laderman, A.D., ed. Atlantic White Cedar Wetlands. Westview Press, Boulder, CO. USA.

Schwintzer, C.R. and Tomberlin, T.J. 1982. Chemical and physical characteristics of shallow groundwaters in northern Michigan bogs, swamps and fens. American Journal of Botany 69:1231-1239.

Solorzano, L. 1969. Determination of ammonia in natural waters by the phenolhypochlorite method. Limnology and Oceanography 14:789-801.

Stone, W. 1911. The plants of southern New Jersey with especial reference to the flora of the Pine Barrens and the geographic distribution of the species Annual report of the State Museum of New Jersey, 1910. Trenton, NJ. USA. 828p.

Swales, S. 1982. Environmental effects of river channel works used in land drainage improvement. Journal of Environmental Management 14:103-26.

CONSEQUENCES OF MANAGEMENT PRACTICES ON COASTAL WETLANDS IN THE EAST AND WEST INDIES[1]

Armando A. de la Cruz
Department of Biological Sciences
Mississippi State University, P.O. Drawer GY
Mississippi State, MS 39762 U.S.A.

ABSTRACT

Coastal wetlands in Southeast Asia and The Caribbean have undergone tremendous cultural alterations. Case studies from the West Indies (Jamaica and Trinidad) and East Indies (Indonesia and Philippines) are used to describe the impacts of management activities. In general, initial management implementation resulted in habitat change, if not degradation, of the wetland environment as exemplified by the Negril Morass in Jamaica and the Caroni Swamp in Trinidad. Subsequent management plans emphasized multiple cultural uses, which incorporate rehabilitation to a semi-natural condition. Indonesia has a model to demonstrate the implementation of this management approach.

[1]Supported in part by grants from the National Science Foundation Division of International Programs (NSF/SEED No. INT-7707133 and NSF No. INT-7920602).

INTRODUCTION

The use of wetlands by humans dates back to ancient times. Remnants of human settlements of swampy areas can still be found today among the Indians of Peru and Bolivia inhabiting the fringes of Lake Titicaca and the Madan tribe of the Tigris-Euphrates reedlands. The dependence of humans on wetlands for thatch and weave materials and for wildlife and fishery food items has served as the basis for the local opposition to destroying swamps. In certain areas, even at the time of mounting pressures, deep-rooted taboos among the local inhabitants has prevented large scale clearance of permanent swamps (Thompson 1983).

In most areas of the world, however, population explosions and agricultural development have encroached upon even the remotest wetlands. Throughout the world, and especially in tropical regions, large scale reclamation is now taking place. Developers armed with management plans often do not really know the ecological and economic consequences that may ensue during the following reclamation development. When the management plan fails, the development is either discontinued or abandoned. There are many examples of attempts and failures to manage wetlands. The aim of this paper is to illustrate examples from the West and East Indies based on personal observations during visits to Trinidad, Jamaica, and Southeast Asia.

Wetland Management Cases

Caroni Swamp (Trinidad). This swamp is located on the west coast of Trinidad, southeast of Port-of-Spain (Fig. 1A). The rivers Caroni and Madame Espagnol drain approximately 670 km^2 of mountainous area and flow through the swamp before entering the Gulf of Paria.

Early in the century, the government of Trinidad and Tobago initiated a reclamation scheme that will drain the major portion of the swamp for agricultural purposes. The reclamation project proceeded in two phases (Bacon 1975).

1) The Savannah Drainage Scheme - The eastern section of the Caroni Swamp called the "Savannah Scheme" was intended to provide land for rice and sugarcane cultivation (Fig. 1B). The area was drained by a series of east-west canals that connected to a north-south main canal. The flow of water was regulated by a North sluice at the Caroni River junction and a South sluice at the Madame Espagnol River junction.

2) The Cipriani Drainage Scheme - The purpose of the "Cipriani Scheme" was to control

D. F. Whigham et al. (eds.), Wetland Ecology and Management: Case Studies, 79–83.

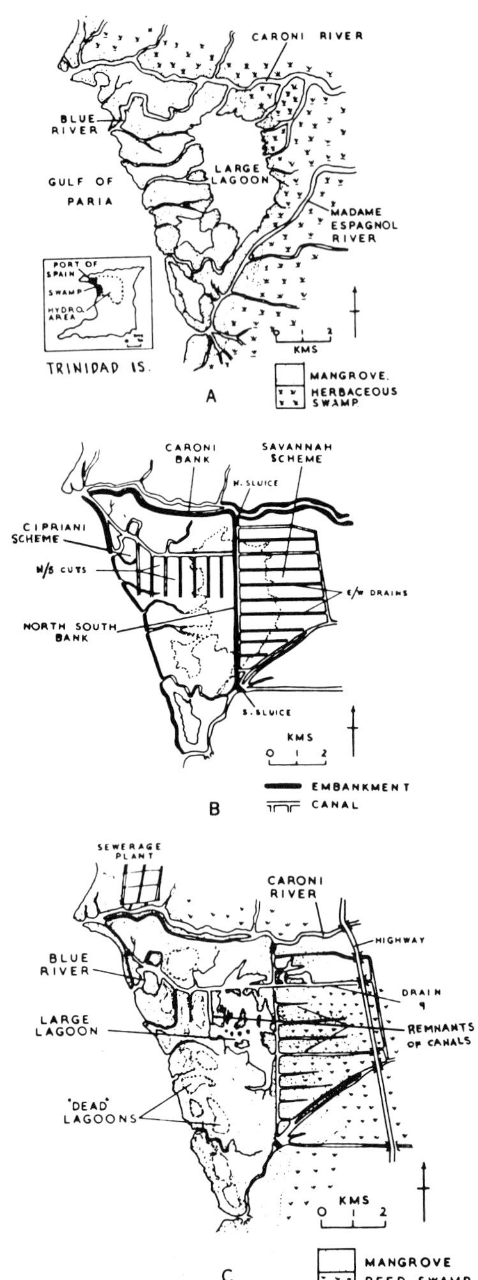

Figure 1. The Caroni Swamp Reclamation Project: A. Prior to reclamation. Insert shows the island of Trinidad and the geographic location of Caroni Swamp in relation to the capital city of Port-of-Spain; B. After development of the Savannah and Cipriani drainage schemes; C. Remnants of the reclamation project after more than half a century of abandonment. (From P.R. Bacon 1975).

flooding caused by the entry of water from the Caroni River into the swamp, and to reclaim by drainage the herbaceous marshes and mangrove swamp areas. The scheme involved channeling the Blue River and excavating 7 north-south cuts from the river (Fig. 1B). These cuts were supposed to increase water circulation and allow drainage from the lagoon and mangrove forest. The water emptied into the Gulf via the mouths of Blue River and Caroni River.

In both schemes, the embankments were built of peat and mangrove trees cleared from the areas. These embankments proved to be too frail against flood water and tides, and breaks occurred allowing salt water into the polders. The operation of the sluice gates were also faulty and their maintenance neglected. The partially reclaimed land underwent natural compaction so that free drainage by gravity was prevented, thus water accumulated and the Savannah section remained inundated for most years.

Today, some 60 years later, the area is again dominated by approximately 4,000 ha of mangrove lagoon and herbaceous swamps. Remnants of the reclamation project are still evident (Fig. 1C). The swamp is again an important habitat for birds and a nursery ground for fishery resources.

Negril Morass (Jamaica). The Negril Morass occupies an area of about 2290 ha within the Negril Basin on the northeast end of the island of Jamaica (Fig. 2). The natural surface drainage pattern in the wetlands has been altered extensively by a network of canal ditches which were constructed to facilitate land reclamation (Natural Resources Conservation Department of Jamaica 1981). The present wetland vegetation consists of Cladium, Rhynchospora, Typha, Acrostichum, with mangroves along the coastline and lowland woodland and swamp forests along the eastern boundary. The boundaries are the North Control Canal on the north, South Control Canal on the south, North/South Canal on the east, and Negril Beach on the west. Fresh water inflow comes from the surrounding hills by way of numerous springs on the eastern edge of the morass.

The reclamation scheme succeeded in converting the wetland east of the North/South Canal into sugarcane field and pastureland but only with the aid of pumping stations. The construction of the canals and diking of the morass partially drained the high marsh resulting in the development of less productive ecotypes of sawgrass and cattails. The levees also isolated the low marsh areas into permanently inundated freshwater wetlands.

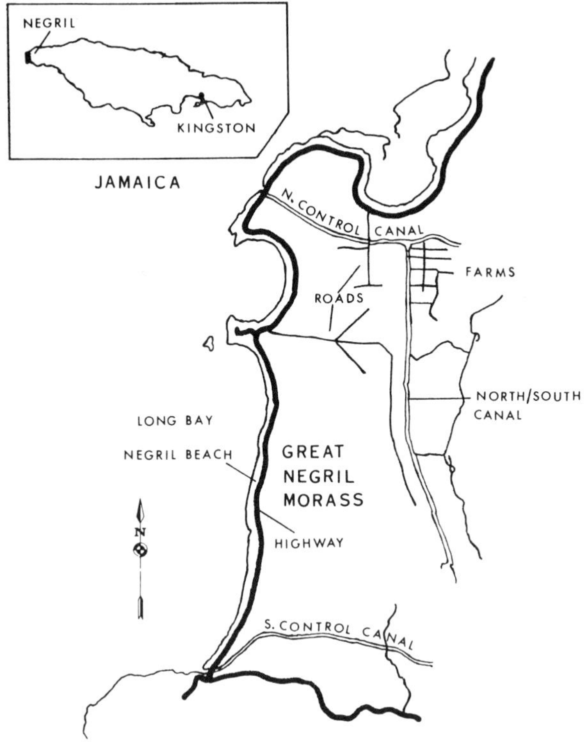

Figure 2. Negril Morass in Jamaica. Insert shows the approximate geographic locaton of the morass on the island. (Modified from NRCP 1981).

Mangrove Swamps (Philippines). Philippine mangrove swamps as well as those of Malaysia, Borneo, and Indonesia suffered massive destruction during the 1960's when mangrove wood had a good market as logs and chips. The mangrove chips were shipped extensively from Southeast Asia to Japan for the production of a viscose-rayon fiber (de la Cruz 1979). High-alpha (dissolving) pulps were prepared by the prehydrolysis sulfate pulping multi-stage bleaching process from the red mangrove Rhizophora (Nicolas and Bawagan 1970). Vast areas of shorelines and whole mangrove islands were completely cleared, became denuded, and silt eroded out into the nearshore waters.

In some coastal villages, shoreline erosion was so bad, houses were threatened. Bulkheads and levees were constructed, usually with Rhizophora pilings cut from a neighboring swamp and coral heads taken from the nearby fringing reefs (de la Cruz 1983). Transplantation of seedlings collected from the field did not work on the high energy side of barriers but was successful in the protected sites. Villagers inhabiting these mangrove areas

Figure 3. Self-initiated tree planting in mangrove areas is done by coastal inhabitants in the Philippines but the practice is localized and limited.

realize the value of mangrove resources to their livelihood. Meager attempts to reforest (Fig. 3) denuded shores and mud banks were generally unsuccessful due to lack of know-how and understanding of the environmental requirements of plant growth. There is a need for a systematic forestry extension service by the national government in mangrove areas.

At present, the alteration of wetlands in Southeast Asia and other third world countries are for aquaculture and agriculture development. The government of these nations promotes wetland development in the name of food production, exportable products, and foreign exchange benefits. It is imperative that wetland management be conducted employing approaches that have been developed for agricultural and forestry management, including some kind of mangrove land reform and conservation practices.

Tambak Rehabilitation (Indonesia). Since World War II, the primary mangrove forests along the coasts of the island of Java have been totally cleared and converted to "tambaks," except for one area in Tjilatjap. Reasons for the massive clearing of mangroves included: (1) harvest of wood for fuel, charcoal-making, lumber products, and other domestic uses; (2) conversion to other land uses such as rice farming, aquaculture ponds for shrimp and milkfish. In the late 1960's, the Indonesian Forest Planning Service and Forestry Department embarked on a management scheme of multi-usage concept of a permanently damaged resource (de le Cruz 1984). Tambak areas are diked into essentially wet forest silviculture where mangrove trees are farmed. Depending on the species, wood biomass is harvested after 3-4 years (Rhizophora). The

Figure 4. Rhizophora plantation in Maura Blanakan, Sukamundi, West Java. The trees are planted from fruits with 2-3 m spacing and harvested 20-30 years depending on preferred size. The 1 ha plots are surrounded by canals which serve as fish ponds.

reforestation programs for idle tambaks are done primarily at shoreline areas most adjacent to tidal influx. The hectare-size plots (Fig. 4) are diked and surrounded by regularly maintained 2-3 m wide canals which served as tidal drainage and natural fish ponds where both Tilapia and Chanos are reared from seed fish without supplementary feeding, fertilizer addition or pesticide application. Fish harvest is done every 4-5 months yielding 200-300 kg/ha for Chanos and 800-1000 kg ha^{-1} for Tilapia. Plots already diked and channelized and awaiting planting are temporarily used as salt ponds during dry season yielding 250 kg d^{-1} 5 m^{-2} subplots; and flooded as fish ponds during the rainy season (de la Cruz 1984).

CONCLUSION

The utilization and management of wetlands in tropical countries will not only continue but it is expected to accelerate as space to grow food and for people to live become increasingly in critical supply. Unmanaged utilization has led to complete destruction of certain coastal wetlands (e.g., in Southeast Asia) while reclamation attempts had either failed or had met minimal success (e.g., in the Caribbean). Hydrological changes caused severe environmental impact. Embankment construction in the Caroni Swamps isolated some areas of mangrove forest from free water circulation which killed the vegetation and led to development of barren pools locally known as "dead lagoons." Channelization and diking of the Negril Morass partially drained the high marsh resulting in the development of less productive stands of sawgrass and cattails. The levees isolated the wetland from intrusion of saline water, thus, the morass is now typically a freshwater marsh in spite of its proximity to the sea. Meanwhile indiscriminate removal of vegetation caused erosion of topsoil making wetland recolonization difficult if not totally impossible, increased water turbidity, loss of habitats, and low estuarine productivity.

ACKNOWLEDGEMENT

I thank the following colleagues who assisted me in conducting field observations and provided me with valuable information: Dr. Eugene Ramcharan of the Institute of Marine Affairs of Trinidad and Tobago; Dr. Peter Reeson and Mr. Donovan Rose of the Natural Resources Conservation Department of Jamaica; Mr. Sukristijono Subardjo, Jan Rachman Hidayat, and Tomi Soemawidjana of Indonesia; Mr. Rastan of the Muara Blanakan Mangrove Reforestation Project in Sukamundi; Dr. Prudy Conlu of the University of Philippines College of Fisheries; and Mr. Cesar Arroyo of the Philippine Mangrove Research Center.

REFERENCES

Bacon, P. R. 1975. Recovery of a Trinididian mangrove swamp from attempted reclamation. pp. 805-815. In Walsh, G; Snedaker, S; and Teas, H. eds. Proceedings of the International Symposium of Ecology and management of Mangroves Vol. II., Institute Food and Agricultural Sciences, University of Florida, Gainesville, FL. USA.

de la Cruz, A. A. 1976. The function of coastal wetlands. Association Southeastern Biologists Bulletin. 23: 179-185.

de la Cruz, A. A. 1979. The functions of mangroves. pp. 125-138. In Srivastava, P.B.L., ed. Mangrove and Estuarine Vegetation of Southeast Asia. Special Publ. No. 10, BIOTROP, Bogor, Indonesia.

de la Cruz, A.A. 1983. Traditional practices applied to the management of mangrove areas in Southeast Asia. International Journal of Ecology and Environmental Science 9:13-19.

de la Cruz, A.A. 1984. A realistic approach to the use and management of mangrove areas in Southeast Asia. pp. 65-68. In Teas, H. J., ed. Physiology and Management of Mangroves Dr. W. Junk Publisher, The Hague, The Netherlands.

Natural Resources Conservation Department. 1981.Environmental Feasibility Study of the Jamaica Peat Resources Utilization Project. Final Technical Report, Vol. 2. NRCD, Ministry of Mining and Energy, Kingston, Jamaica. 413 p.

Nicolas, P.M. and Bawagan, B. D. 1970. Production of high-alpha (dissolving) pulps from Bakawan-babae (Rhizophora mucronata Lam.). The Philippine Lumberman. 164 p.

Thompson, K. and Hamilton, A. C. 1983. Peatland and Swamps of the African Continent. pp. 331-373. In Gore, A. J. P., ed. Mires, Swamps, Bog, Fen and Moor. 4B Regional Studies Elsevier Scientific Publishing Company, Amsterdam, The Netherlands.

98

EFFECT OF PEATLAND ON WATER QUALITY, FISH AND WILDLIFE HABITAT IN CANADA, A REVIEW

V. Glooschenko
Wetland Habitat Coordinator
Wildlife Branch Ontario Ministry of Natural Resources
Whitney Block
Toronto, Canada M7A 1W3

ABSTRACT

Peatland drainage can influence the quality of receiving waters with respect to increased sedimentation, nutrient limitation, dissolved oxygen and organic carbon and possible release of mercury with its subsequent accumulation in fish and other biota. Development of any kind has a profound effect on vegetation patterns and on wildlife inhabiting peatland areas. The conclusion drawn by a number of workers is that the environmental balance of peatlands appears to be so sensitive that development permanently destroys peatland ecosystems.

INTRODUCTION

Peatland ecology in North America has received little attention from either academic or applied ecologists. Information is surprisingly lacking on many of the fundamental features of peatland ecology, the interaction of environmental variables, and individual life-history differences that control peatland organisms (Gorham 1982). Keddy (1981) has stated that not only is there a lack of understanding of the physical factors which produce a given wetland type but also how these factors influence most of the basic processes occurring in wetlands. This limited theoretical and practical knowledge of peatland ecology has obvious implications for environmental impact assessment (Washburn and Gillis 1982).

However, we know that habitat plays a crucial role in the survival of peatland flora and fauna. Plant species dependent on peatlands are especially adapted to conditions of low pH, high organic content, and a waterlogged, anaerobic environment with slow decomposition due to low microbial activity. Such adaptations restrict these organisms to peatland areas.

Specialized small mammals feed on abundant peatland invertebrates and are themselves key prey in the peatland food chain. These organisms are restricted to peatland areas. In addition, major groups of birds including grouse, migratory waterfowl, and songbirds depend on peatlands either year round or on a seasonal basis.

The environmental impacts resulting from peat mining development are investigated in this report with respect to effects on water quality, vegetation and wildlife habitat. Examples have been drawn to reflect the Ontario situation where possible. Ontario's 16,000,000 hectares of peatland (Monenco Ontario Limited, 1981) range from the northern tundra to the southern deciduous forest zone. They include bogs, fens, swamps and marshes each with characteristic ecological systems, and each dependent on climate, water flow, and nutrient content.

WATER QUALITY

pH and Temperature

Although Clausen and Brooks (1983) observed an increase in acidity (as $CaCO_3$), mean values for pH remained unchanged between peat mined and control bogs over their two year study. Moore (1987) found no difference in pH between disturbed and undisturbed bogs even immediately following ditching. Other workers have noted an increased pH in runoff from several mined Finnish peatlands compared to undisturbed bogs (Sallantaus and Patila 1983) and from peatland forest drainage (Ramberg 1981). Increased pH in receiving waters from peatland drainage does not appear to be of concern.

However, there are two primary environmental impacts on receiving water associated with the pre-production phase of peatland development. These are: local increases in water temperature and an increase in organic sediment loads.

D. F. Whigham et al. (eds.), Wetland Ecology and Management: Case Studies, 85–93.

During the summer, the lack of shade from vegetation allows solar heating to increase soil temperature much higher than that of the surrounding terrain. The temperature increase is passed on by heat conduction to the water in the peat, in turn raising the temperature of the water draining from the peatland. Clausen and Brooks (1983) found water temperatures of a mined bog to be 5^0 warmer than water draining from control bogs. The release of this water into cooler stream water may create a thermoclinetic barrier to migrating fish, as well as affecting other temperature sensitive aquatic species. Studies have shown that increases in water temperature can also increase levels of methyl mercury released from sediments (Rudd et al. 1983; Parks et al. 1984).

Dissolved Oxygen Levels

The percent oxygen saturation of drainage water may be higher during mining activities. This could be caused by higher water temperatures and not by an increase in dissolved oxygen concentration (Clausen and Brooks 1983).

However, the same authors note that dissolved oxygen concentration in peatland drainage water may be reduced by mining activities. This is due to increased organic matter in the water which depletes dissolved oxygen levels as it decomposes. Discharge rates from peatlands affect dissolved oxygen levels, since turbulence created during high flow introduces more atmospheric oxygen into the water. When the flow slows, suspended peat particles fall out of suspension and continue to decompose. If there is a large concentration of settled organic matter, dissolved oxygen concentrations can become low enough to cause an oxygen depression. Oxygen depression can present a barrier to migrating fish, or damage eggs of fish and other organisms.

For management purposes, dissolved oxygen levels may be increased by either introducing submergent plant species into the outlet pond, or construction of low waterfalls at the outlet if elevation permits.

Organic Loading

Bourbonniere (1987) found that dissolved organic carbon levels were elevated by approximately 30% in water from drained bogs, increasing the biochemical oxygen demand by about 15%. However, Moore (1987) found dissolved organic carbon levels were not affected by disturbance except for a decline in one bog from which peat had been removed down to the underlying sand, and a short-term increase immediately following ditching in another bog. Although there may be no change in runoff water quality, the increased runoff following drainage is likely to impose a greater organic loading on the receiving system (Moore 1987).

The flux of organic material from peatlands could have significant effects on the watershed fish populations. Maximum organic carbon levels were found in summer (Bourbonniere 1987), coinciding with low oxygen capacity and high oxygen demand in streams. Decomposition of the organic matter would reduce dissolved oxygen levels in the stream causing oxygen deficiencies (Clausen and Brooks 1983), hastening eutrophication, releasing large quantities of nutrients (Clausen 1980) enabling algal populations to thrive, and creating problems for fish populations. The addition of organic material has also been shown to increase methyl mercury production in mercury-polluted sediments.

Nutrient Levels

Levels of organic nutrients have been shown to increase with peat mining activities (Clausen and Brooks 1983; Moore 1987). The release of nutrients is related to increased rate of decomposition. Drained peatlands provide greater aeration and warmer soil temperatures that permit micoorganisms to decompose peat at a faster rate (Clausen 1980). Nutrients released in the largest quantities are nitrogen and phosphorus which can be responsible for accelerating eutrophication in receiving waters. Nutrient levels fluctuate annually, with low release levels occurring in spring and a maximum release level in winter under ice conditions (Clausen 1980) or in late fall, just prior to freezing, due to death of plants (Richardson and Marshall 1986). Maximum nutrient release from drained peat would be expected to occur in summer when microbial activity is at its peak. The apparent lag in nutrient release is probably due to the slow rate of water transport through peatlands and rapid removal of nutrients by microorganisms (primarily fungi and yeasts) and physical sorption, as shown for phosphorus by Richardson and Marshall (1986). The latter authors also found that freezing of peat resulted in phosphorus release, presumably due to reduction of biological uptake, interference with sorptive mechanisms, and possible lysis of microorganisms. Nutrient levels in drainage water can be reduced by introducing pond vegetation at ditch outlets which can utilize some of the released nutrients (Hazen and Beeson 1979). The winter maximum for nutrient release would be difficult to eliminate.

Humic waters from peat drainage may also reduce productivity due to nutrient limitation. Jackson and Schindler (1975) found that humic complexes of Al and Fe could bind phosphate ions thus lowering the bioavailability of P. Richardson (1985) showed that the ability of soils from a variety of wetlands to retain added phosphorus could be predicted solely from the extractable aluminum content of the soil. Jackson and Hecky (1980) noted that primary productivity of lakes and reservoirs in northern Manitoba was inversely proportional to DOC and non-dialyzable Fe, organic carbon, nitrogen and H-bonded polymeric hydroxyl groups. This would suggest allochthonous humic-FeOH complexes with strong, covalent metal-ligand bonds. These workers concluded that primary production was not depressed by lowered pH, reduced light or bioavailability of P; but they suggested that Fe and other essential micronutrients such as metals could be sequestered by formation of colloidal humic - Fe complexes. Also, some binding of P could take place.

Suspended Sediment Levels

A significant environmental impact may be associated with the release of organic peat particles eroded from exposed peat surfaces.

Variability in suspended sediment transport from mined peatlands has been observed by a number of workers. Sallantaus and Patila (1983) observed the main portion of suspended sediment loading from mined peatlands to occur during the ditch excavation period with the period of snow melt. Although there was a decrease in loading during the initial year after ditching, suspended solid levels (<10mg/l) were still in excess of those from unmined peatlands (<1mg/l). Heikurainen et al., (1978) also observed peaking of suspended solid loading after ditching. Korpijaako and Pheeney (1976) did not note an increase of suspended peat sediment in peatland drainage waters. Mundale (1981) however noted a large increase in suspended sediments at the initiation of drainage with a general reduction over time and and periodic rises following storm events.

Clausen and Brooks (1983) found mean suspended sediment levels more than twice the levels associated with control bogs over a two year period. Stewart and Lance (1983) measured suspended sediment loads in drained peatlands up to ten times higher during baseflow and one hundred fold higher during storms than before mining activities. Immediately after draining, they measured sediment loads after a twenty- four hour storm equal to the yearly peatland sediment output prior to drainage. In the first year after drainage the mean suspended sediment level was 17 times that of pre- drainage. Six years later the levels were still three times higher than pre-drainage.

When considerable erosion of mineral soil has occurred, discharge of suspended sediments up to several tons per km^2 have been observed in forest drainage areas (Seuna 1980).

An increase in suspended sediments causes increased turbidity which may reduce photosynthesis and thus the availability of plant food. The accumulation of sediment on the stream bed decreases survival of eggs and alevins, as well as killing bottom vegetation which is a major food source for fish and aquatic invertebrates (Hill 1976). Some management of sediment loading may be achieved by ditching in the winter (Sallantaus and Patila 1983) or early summer (Heinkurainen et al. 1978) and by creation of settling ponds at the outlet of the perimeter drainage ditches the reduce both discharge and suspended sediment levels during storm events (Hazen and Beeson 1979).

Influence of Peatland Drainage Water on Mercury Release

Increased discharge of Hg related to peatland drainage is of environmental concern. Simola and Lodenius (1982) studied the vertical distribution of Hg in a core taken from a Finnish lake. They noted that Hg doubled in concentration starting in the late 1960's with a flux rate that increased five times in the period 1967-1977 compared to the end of the 19th century. Peatland drainage was considered the cause of the increased Hg loading since no agricultural or industrial sources were present. Increase of solubility of Hg due to increased dissolved organic carbon was the proposed mechanism. Jackson et al. (1980) had previously noted increased solubility of Hg due to the formation of stable organic complexes. Lodenius (1983) found increased Hg in pike in lakes in Finland which received peatland drainage. He related the Hg increase to several environmental factors including low pH and high humic matter. In a subsequent study Lodenius et al. (1983) noted that the transport of Hg from peat soil was facilitated by humus. Rudd et al. (1983) also observed that addition of natural organic matter, equally Sphagnum moss, can increase the methylation of Hg in the water column.

For management purposes, care should be taken to reduce or eliminate peaty organic material or sphagnum from entering receiving waters where mercury release from sediments may be of concern (Glooschenko et al. 1984).

INFLUENCE OF LOWERED WATER TABLE ON BIOTA

A change in water table levels has a profound effect on the biotic composition of the drained area (Keddy 1981). Drawdown to a level below root penetration causes death to any remaining vegetation and displaces wildlife dependent on the plant life. The effect on the surrounding terrain is seen as a gradual, off-site change in species composition reflecting changes in vegetation patterns resulting from alteration in hydrology (JRB Associates, Inc. 1981). An example of this is the succession of Picea-Larix forest to a Betula-Populus or Picea-Abies association. Each of these forest types will attract and support different wildlife systems. Ruuhijarvi (1972) noted a considerable decrease in bird numbers, especially in wading birds, ducks and gamebirds, as an area is drained.

Alteration to Plant Species Composition

Many workers have noted relationships between the composition of wetland communities and water table fluctuations (Conway 1949; Pearsall 1950; Jeglum 1973;1974; Vitt and Slack 1975) as well as with water nutrient levels (Heinselman 1970; Jeglum 1971; Sparling 1979; Riley 1982; Tallis 1983).

Sjors (1980) has noted that the greatest ecological impact from peatland drainage will come from the reduction of specialized plants as well as of animals. A ditch dug in 1938 in the eastern section of the Mer Bleue bog near Ottawa lowered water levels permitting the invasion of a variety of plants usually foreign to this type of bog including trembling aspen (Populus tremuloides) and balsam poplar (Populus balsamifera), (Joyal 1970). In northern Michigan, clear-cutting and ditching have resulted in the replacement of coniferous bog forests by red maple (Acer rubrum) and alder (Alnus spp.) in some areas (LeBarron and Neetzel 1942). In Ontario, rich fens are considered the most threatened of all wetlands; it has been recommended (Sparling 1979) that they be spared exploitation in favour of less significant areas.

Provincially Rare Peatland Flora

Orchids and other rare plants are being extirpated from much of their former range particularly in Southern Ontario. A number of rare flora (Argus and White 1977) are restricted to peatlands which may be subject to peat drainage and mining. Some of these are the white fringed orchid (Platanthera blephariglottis), yellow fringed orchid (Platanthera ciliaris), prairie white fringed orchid (Platanthera leucophaea), southern twayblade (Listera australis), yellow eyed grass (Xyris difformis) and arrow arum (Peltandra virginica). The small white ladies slipper (Cypripedium candidum) also found in this restricted habitat is an endangered species.

Peatland Birds

Many species of birds are attracted by berries and seeds found in peatland habitats as well as by an abundance of insects and other small invertebrates. Common bird species which nest in forests near bog areas are solitary vireo (Vireo solitarius), ruby-crowned kinglet (Regulus calendula), solitary sandpiper (Tringa solitaria), black-capped chickadee (Parus atricapillus), dark-eyed junco (Junco hyemalis), and Tennessee (Vermivora peregrina) and Nashville warblers (V. ruficapilla) (Godfrey 1966).

Flycatchers, wood warblers and several sparrow species are found in numbers in wooded peatlands and edge habitats of subarctic, boreal and temperate peatlands. These species are migratory and utilize peatlands for breeding.

A variety of ducks and geese as well as tundra swans seek bogs, fens, and open water habitat for nesting, rearing, feeding and staging.

Removal of vegetative cover produces an open expanse which is unattractive to waterfowl (Montreal Engineering Company Limited 1978) possibly upsetting migrating patterns, nesting and rearing habits. Ultimately, drainage and lowering of the water table will cause much former habitat to become unusable to waterfowl and other species.

Spruce grouse (Canachites canadensis) require black spruce (Picea mariana) thickets found in association with wooden bogs and swamps while ruffed grouse (Bonasa umbellus) and willow ptarmigan (Lagopus lagopus) seek more open areas with shrub cover (Crichton 1963; Anderson 1973; Marshal and Miquell 1978).

The predatory short-eared owl (Asio flammeus) and northern harrier or marsh hawk (Circus cyaneus) are typically found in fens, where they take rodents and to a lesser extent, birds and fish.

These and other birds commonly

Table 1. Some birds commonly associated with Canadian peatlands (from Clarke- Whistler and Rowsell 1982).

Legend:
Status (I): E = rare, endangered, threatened; U = uncommon; C = common
Peatland Types (II): B = bog; F = fen; S = swamp; E = edge; O = open water
Relative Importance to User (III): C = critical; S = important (seasonal); M = marginal

SPECIES	(I)	(II)	(III)
Common Loon	U	O	S
American Bittern	C	F,E	S
Tundra Swan	E	F,O	C
Canada Goose	C	B,F	C
Mallard/Black Duck	C	B,F	S
Ring-necked Duck	C	B	S
Common Merganser	C	S,O	S
Northern Harrier	C-U	F	S
Osprey	U	B,O	C
Gyrfalcon	E	B,F	M-S
Peregrine Falcon	E	B,F	M-S
Spruce Grouse	C	B	C
Ruffed Grouse	C	B,E	S
Sharp-Tailed Grouse	C	F,E	S-C
Whooping Crane	E	F	C
Sandhill Crane	U-E	B,F	S-C
Yellow Rail	U	F	S
Common Snipe	C	B,F	S
Eskimo Curlew	E	B,F,E	C
Solitary Sandpiper	C	B,F	S
Greater Yellowlegs	C	B,F	S
Lesser Yellowlegs	C	B,F	S
Bonaparte's Gull	C-U	F	S
Short-eared Owl	U	F	S
Yellow-bellied Sapsucker	C	S,E	S
Black-backed Woodpecker	U	S,E	M-S
Eastern Kingbird	C	B,F,S	M-S

Table 1. Continued.

Legend:
Status (I): E = rare, endangered, threatened; U = uncommon; C = common
Peatland Types (II): B = bog; F = fen; S = swamp; E = edge; O = open water
Relative Importance to User (III): C = critical; S = important (seasonal); M = marginal

SPECIES	(I)	(II)	(III)
Yellow-bellied Flycatcher	C	B	S
Alder Flycatcher	C	B,F	S
Tree Swallow	C	E	S
Gray Jay	C	B,S	M-S
Boreal Chickadee	C	B,S,E	S
Sedge Wren	U	F	S
Hermit Thrush	C	B	S
Ruby-crowned Kinglet	C	B,S	S
Solitary Vireo	C	B,S	S
Tennessee Warbler	C	B,S	S
Nashville Warbler	C	S,E	S
Yellow Warbler	C	B,S,E	S
Palm Warbler	C	B	S
Northern Waterthrush	C	S,E	S
Common Yellowthroat	C	B,F,S	S
Wilson's Warbler	C	B,S,E	S
Red-winged Blackbird	C	B,F,S,E	S
Rusty Blackbird	C	B,S	S
Savannah Sparrow	C	B,F	S
Le Conte's Sparrow	C	B,F	S
Swamp Sparrow	C	B,S	S
Song Sparrow	C	B,S,E	M-S

associated with Canadian peatlands are listed on Table 1, together with indication of the relative importance of peatland to the species (Clarke-Whistler and Rowsell 1982). For 35 species listed peatland habitat is seasonally important and for 8 species the habitat is of critical importance.

Peatland Mammals

Peat harvest causing habitat change or alteration in vegetation communities will particularly affect small mammals with home ranges restricted primarily to peatlands.

Shrews, moles, rodents, hares and rabbits are the major small mammal group found in peatlands. Many are important food sources for avian and mammalian predators and have high reproductive rates. Gapper's red-backed vole (*Clethrionomys gapperi*) and both the southern and northern bog lemming (*Synaptomyus cooperi*, *S. borealis*) are found in open bogs and fens. Bog lemmings appear to be the most restrictive of any mammals using peatlands, as they are found primarily in *Carex* sedges and grasses (Connor 1959). These animals are key prey items for a number of predators including lynx (*Felis lynx*), short-eared owl (*Asio*

Table 2. Some mammals commonly associated with Canadian Peatlands (from Clarke-Whistler, Rowsell 1982).

Legend:
Status (I): E = rare, endangered, threatened; U = uncommon; C = common
Peatland Types (II): B = bog; F = fen; S = swamp; E = edge; O = open water
Relative Importance to User (III): C = critical; S = important (seasonal); M = marginal

SPECIES	(I)	(II)	(III)
Small and Medium Mammals			
Masked Shrew	C	B,F,S,E	C
Vagrant Shrew	C	B,F,S,E	C
Arctic Shrew	C	B,F,S,E	C
Pigmy Shrew	C-U	B,F,S,E	S-C
Short-tailed Shrew	C	B,F,S,E	S
Star-nosed Mole	C	B,F,S,E	S-C
Snowshoe Hare	C	B,F,S,E	S
Arctic Hare	C-U	B,F	M
Least Chipmunk	C	S,E	M
American Red Squirrel	C	S	S
Northern Flying Squirrel	C	S	S
American Beaver	C	B,F,S,E,O	S
Deer Mouse	C	S,E	M
Gapper's Red-backed Vole	C	B,F,S,E	C
Southern Bog Lemming	C	B,F,S,E	C
Northern Bog Lemming	C	B,F,S,E	C
Heather Vole	C	B,F,S,E	M-S
Muskrat	C	F,S,E,O	S
Meadow Vole	C	B,F,E	S
Chestnut-cheeked Vole	U	B,F,S,E	M-S
Meadow Jumping Mouse	C	B,F,S,E	S
Woodland Jumping Mouse	C	S,E	M
American Porcupine	C	S	M

Table 2. Continued.

Legend:
Status (I): E = rare, endangered, threatened; U = uncommon; C = common
Peatland Types (II): B = bog; F = fen; S = swamp; E = edge; O = open water
Relative Importance to User (III): C = critical; S = important (seasonal); M = marginal

Species	(I)	(II)	(III)
Bats			
Little Brown Bat	C	S,E	M
Hoary Bat	U	S,E	M
Carnivores			
Coyote	C	S,E	M
Wolf	C	B,F,S,E	M
Red Fox	C	S,E	M
American Black Bear	C	B,F,S,E	M-S
Raccoon	C	S,E	S
American Marten	C	S,E	S
Fisher	C-U	S,E	S
Ermine	C	S,E	S
Least Weasel	C	B,F,S,E	M
American Mink	C	B,F,S,E	M
River Otter	C	E,O	M
Lynx	C	S,E	S
Ungulates			
Caribou	C	B,F,S	S-C
White-tailed Deer	C	F,S,E	M-S
Moose	C	B,F,S,E	S
American Bison	U-E	F,S	S

flammeus) and northern harrier (Circus cyaneus). The masked shrew (Sorex cinereus) and arctic shrew (Sorex arcticus) are most abundant in open peatlands where small insect numbers are high (Getz 1961); the short-tailed shrew (Blarina brevicauda) is more often seen in hardwood swamps and fens (Buckner 1966). Prior to harvest, peatland drainage will alter the quality of ground-water as well as its collection and flow patterns (Mundale 1981). This in turn changes growth patterns of vegetation which affect food sources and cover for wildlife. A lowered groundwater level would be especially detrimental to the specialized northern and southern bog lemming and Gapper's red-backed vole. The snowshoe hare (Lepus americanus) relies on coniferous swamp, swamp edge and fen habitat (Keith 1966; Marshal and Miquelle 1978) and is prey for many predators. Red squirrel (Tamiasciurus hudsonicus), least chipmunk (Eutamias minimus), and northern flying squirrel (Glaucomys sabrinus) use upland, mesic, and swamp peatland although their primary habitat is coniferous forest. Bogs support caribou lichens and form an important habitat type in some parts of woodland caribou (Rangifer caribou) range. Moose (Alces alces) frequently feed and calve in swamps and peatland edge. Fens are used within a restricted range by american bison (Bison bison) and muskox (Oribos moschatus) (Banfield 1974, Wiken et al. 1981).

Table 2 shows these and other mammals associated with Canadian peatlands; for 15 of these species peatland habitat is seasonally important and for 9 species it is of seasonal and/or critical importance (Clarke-Whistler and Rowsell 1982).

Amphibians and Reptiles

Amphibians play an important role in the aquatic food chain even if in some peatlands they may be unable to breed successfully (Blanches and McNicol 1986). These workers observed green frogs (Rana clamitans), mink frogs (Rana septentrionalis) and bullfrogs (Rana catesbeiana) to inhabit the pool edge of peatlands in Northern Ontario while toads (Bufo americanus), spring peepers (Hyla crucifer), wood frogs (Rana sylvatica) and leopard frogs (Rana pipiens) were found on the Sphagnum mat.

Two rare and provincially significant species known to inhabit Canadian peatlands are the eastern massasauga rattlesnake (Sistrurus catenatus catenatus) and the spotted turtle (Clemmys guttata). Cook et al. (1980) report that the spotted turtle occurs in bogs in Central Ontario. The eastern massasauga rattlesnake still may be found in reduced numbers in Wainfleet Bog, a southern Ontario peatland currently being drained and harvested and in more secure peatland habitat in Georgian Bay National Park.

Peatland drainage and harvest would completely eliminate reptiles and amphibians along with their food source of aquatic invertebrates.

Fish

Although fish populations of open waters and drainage creeks within peatland show low diversity, northern lakes and rivers with associated peatland drainage areas support fish populations which are significant economically and ecologically. Aquatic habitats with peatland drainage have produced record size fish including grayling (Thymallus arcticus), lake trout (Salvelinus namaycush), arctic char (Salvelinus alpinus), northern pike (Esox lucius) and whitefish (Coregonus clupeaformis) (Scott, Crossman 1969). Lakes and rivers in boreal and temperate peatlands provide habitat for spawning, feeding and overwintering for lake trout, rainbow trout (Salmo gairdnerii), brook trout (Salvelinus fontinalis) and several species of salmon.

ACKNOWLEDGEMENTS

Dr. John Riley, Regional Ecologist with the Ontario Ministry of Natural Resources at Richmond Hill and Mr. Alex Caron, Wildlife Branch, Ontario Ministry of Natural Resources provided advice and references on rare peatland orchids. Mr. John Archbold assisted in reviewing the literature. Their assistance is gratefully acknowledged.

REFERENCES

Anderson, J.L. 1973. Habitat Use, Behaviour, Territoriality and Movements of the Maple Spruce Grouse of Northern Minnesota. M.S. Thesis, University of Minnesota, St. Paul, Minnesota. 119 p.

Argus, G.W. and White, D.J. 1977. The rare vascular plants of Ontario. National Museum of Natural Sciences, National Museums of Canada, Ottawa, Canada 129 p.

Banfield, A.W.F. 1974. The mammals of Canada. National Museum of Natural Sciences, University of Toronto Press, Toronto, Ontario, Canada 438 p.

Blancher, P.J. and McNicol, D.K. 1986. Investigations into the Effects of Acid Precipitation on Wetland-dwelling Wildlife in Northeastern Ontario. Technical Report Series No. 2. Ontario Region Canadian Wildlife Service, Ottawa, Canada 153 p.

Bourbonniere, R.A. 1987. Organic Geochemistry of Bog Drainage Waters. pp. 139-145. In C.D.A. Rubec and R.D. Overend, compilers. Proceedings of Symposium 87 - Wetlands and peatlands. Edmonton, Alberta. August 23-27, 1987 Environment Canada. National Water Research Institute. Burlington, Canada NWRI Contribution #87-133.

Buckner, C.H. 1966. Populations and ecological relationships of shrews in tamarack bogs of southeastern Manitoba. Journal of Mammalogy 47:181-194.

Clarke-Whistler, K. and Rowsell, J.A. 1982. Peatlands as Fish and Wildlife Habitat. In: Sheppard, J.D., J. Musial and T.E. Tibbetts, eds. A symposium on peat and peatlands, Symposium '82. Shippagan, New Brunswick, September 12-14. 582 p.

Clausen, C. 1980. The Quality of Runoff from Natural and Disturbed Minnesota Peatlands. International Peat Society. pp. 523-537. In Proceedings of the 6th International Peat Congress, August 17-23, 1980. Duluth, Minnesota, USA.

Clausen, J.C. and Brook, K.N. 1983. Quality of runoff from Minnesota peatlands: II. A method of of assessing mining impacts. Water Resources Bulletin 19:769-772.

Cook, F.R., LaFontaine, J.D., Black, S., Luciuk, L. and Lindsay, R.V. 1980. Spotted Turtles (Clemmys guttata) in eastern Ontario and adjacent Quebec. Canadian Field-Naturalist 94(4):411-415.

Connor, P.F. 1959. The Bog Lemming

(Synaptomus cooperi) in Southern New Jersey. Michigan State University Museum Publications. The Museum of Michigan State University. East Lansing, Michigan, USA. Biological Series Vol. 1 (5): 166-248.

Conway, V.M. 1949. The bogs of central Minnesota. Ecological Monographs 19: 174-206.

Crichton, V. 1963. Autumn and winter foods of the spruce grouse in central Ontario. Journal of Wildlife Management 27:597.

Getz, L.L. 1961. Factors influencing the local distribution of shrews, American Midland Naturalist 5:67-88.

Glooschenko, V., Service, M., Allen, G., Betts, B., Parker, B. and Dawson, J. 1984. Impact of Peatland Development on Fish and Wildlife Habitat - Literature Review and Recommendations for Impact Assessment in the Dryden District. Technical Report, Wildlife Branch, Ontario Ministry of Natural Resources. Toronto, Ontario, Canada. 102 p.

Godfrey, W.E. 1966. Birds of Canada. National Museum of Natural Sciences, National Museums of Canada,Bulletin 203. Ottawa, Ontario, Canada. 428 p.

Gorham, E. 1982. Some unsolved problems in peatland ecology. Le Naturaliste Canadien 109:533-541.

Hazen, C.B. and Beeson, P.J. 1979. Techniques for the assessment of bog hydrology, harvesting, and mitigation of water quality impacts resulting from peat harvesting. pp. 219-240. In Feingold, B. F, McGrew, W., Masterton, J.K., Farrell, M.D. and White,T. W., eds. Management Assessment of Peat as an Energy Resource, Executive Conference Proceedings. Arlington, Virginia, USA. July 22-24, 1979.

Heikurainen, L., Kenttamies, K. and Laine, J. 1978. The environmental effects of forest drainage. Suo 29:49-58.

Heinselman, M. 1970. Landscape evolution, peatland types, and the environment in the Lake Agassiz Peatlands Natural Areas, Minnesota. Ecological Monographs 40: 235-261.

Hill, A.R. 1976. The environmental impacts of agricultural land drainage. Journal of Environmental Management 4:251-274.

Jackson, T.A. and Schindler, D.W. 1975. The biogeochemistry of phosphorus in an experimental lake environment: evidence for the formation of humic-metal-phosphate complexes. Proceedings of the International Association of Theoretical and Applied Limnology 19:211-221.

Jackson, T.A. and Hecky, R.E. 1980. Depression of primary productivity by humic matter in lake and reservoir waters of the boreal forest zone. Canadian Journal of Fisheries and Aquatic Sciences 37:2300-2317.

Jackson, T.A., Kipphut, G., Hesslein, R.H. and Schindler, D.W. 1980. Experimental study of trace metal chemistry in soft-water lakes at different pH levels. Canadian Journal of Fisheries and Aquatic Sciences 37:387-402.

Jeglum, J.K. 1971. Plant indicators of pH and water level in peatlands at Candle Lake, Saskatchewan. Canadian Journal of Botany 49:1661-1676.

Jeglum, J.K. 1973. Boreal forest wetlands, near Candle Lake, central Saskatchewan, Part II: Relationships of vegetation variation to major environmental gradients. Musk-Ox 12: 32-48.

Jeglum, J.K. 1974. Relative influence of moisture-aeration and nutrients on vegetation and black spruce growth in northern Ontario. Canadian Journal Forest Research 4:114,126.

Joyal, R. 1970. Description de la bourbiere a sphaignes Mer Bleue pres d'Ottawa: I. Vegetation. Canadian Journal of Botany 48: 1405-1418.

JRB Associates, Inc. 1981. Peat Mining: An Initial Assessment of Wetlands Impacts and Measures to Mitigate Adverse Effects. Prepared for the Environmental Protection Agency, Washington, DC. USA. 61 p.

Keddy, P.A. 1981. Biological considerations in wetlands management. pp. 183-189. In A. Champagne, ed. Proceedings of the Ontario Wetlands Conference, 1981. Toronto, Ontario, Canada. 193 p.

Keith, L.B. 1966. Habitat vacancy during a snowshoe hare decline. Journal of Wildlife Management 30:828-832.

Korpijaako, E. and Pheeney, P. 1976. Transport of peat sediment by drainage system from exploited peatland. pp. 135-138 In Proceedings of the 5th International Peat Congress, Warsaw, Poland.

LeBarron, R.K. and Neetzel, J.R. 1942. Drainage of forested swamps. Ecology 23:457-465.

Lodenius, M. 1983. The effects of peatland drainage on the mercury contents of fish. Suo 34:21-24.

Lodenius, M., Seppanen, A. and Uusi-Rauva, A. 1983. Sorption and mobilization of mercury in peat soil. Chemosphere 12:1575-1581.

Marshall, H.M. and Miquelle, D.G. 1978. Terrestrial Wildlife in Minnesota Peatlands. Minnesota Department of Natural Resources. St. Paul, Minnesota, USA. 193 p.

Monenco Ontario Ltd. 1981. Evaluation of the Potential of Peat in Ontario. Ontario

Ministry of Natural Resources, Mineral Resources Branch, Occasional Paper #7. Toronto, Ontario, Canada. 193 p.

Montreal Engineering Company Limited. 1978. The mining of peat - a Canadian energy resource. EMR CANMET Contract Report 7-9047, Ottawa, Ontario, Canada.

Moore, T.R. 1987. A preliminary study of the effects of drainage and harvesting on water quality in ombrotrophic bogs near Sept-Iles, Quebec. Water Resources Bulletin 23: 785-791.

Mundale, S.H. (ed.) 1981. Energy from Peatlands: Options and Impacts. A report of the CURA Peat Policy Project #9. Publications No. 81-2.

Osborne, J.M. 1982. Potential environmental impacts of peatland development. pp. 198-219. In: Sheppard, J.D., J. Musial and T. E. Tibbetts, eds. A Symposium on Peat and Peatlands, Symposium '82. Shippagan, New Brunswick, September 12-15, 1982. National Research Council of Canada, Division of Energy Report 24134. Halifax, Nova Scotia, Canada.

Parks, J.W., Sutton, J.A. and Hollinger, J.D. 1984. Mercury contamination of the Wabigoon/English/Winnipeg River system, 1980 - causes, effects and selected remedial measures. pp. 1-352. In Allan, R.J., and Brydes T., eds. Mercury pollution in the Wabigoon-English River system of Northwestern Ontario, and possible remedial measures. Final Report of the Steering Committee 1:18 pp; and 2:538 p. Ottawa, Ontario, Canada.

Pearsall, W.H. 1950. Mountains and Moorlands. Collins. London, England. 312 p.

Ramberg, L. 1981. Increase in stream pH after forest drainage. Ambio 10:34-35.

Richardson, C.J. 1985. Mechanisms controlling phosphorus retention capacity in freshwater wetlands. Science 228:1424-1427.

Richardson, C.J. and Marshall, P.E. 1986. Processes controlling movement, storage, and export of phosphorus in a fen peatland. Ecological Monographs 56:279-302.

Riley, J.L. 1982. Hudson Bay Lowland floristic inventory, wetlands catalogue and conservation strategy. Le Naturaliste Canadien 109:543-555.

Rudd, J.W.M., Turner, M.A., Furutani, A., Swick, A.L. and Towsend, B.E. 1983. The English-Wabigoon River system: I. A. synthesis of recent research with a view towards mercury amelioration. Canadian Journal of Fisheries and Aquatic Science 40: 2206-2217.

Ruuhijarvi, R. 1972. Multiple use of peatlands in Finland with secial reference to their conservation. pp. 191-202. In: Proceedings of the IV International Peat Congress, Otaniemi, Helsinki, Finland, June 25-30, 1972.

Sallantaus, T. and Patila, A. 1983. Runoff and water quality in peatland drainage areas. p. 183-202. In International Peat Society, Commission III, Proceedings of the International Symposium on Forest Drainage, Tallinn, U.S.S.R., September 19-23, 1983. Helsinki, Finland: IPS Commission III.

Scott, W.B. and Crossman, E.J. 1973. Freshwater fishes of Canada. Bulletin 184, Fisheries Research Board of Canada, Ottawa, Canada. 966 p.

Seuna, P. 1980. Long-term influence of forestry drainage on the hydrology of an open bog in Finland. The influence of man on the hydrological regime with special reference to representative and experimental basins. Proc. of the Helsinki Symposium 23-26 June 1980. IAHS-AISH Publ. 130, 141-150. IAHS Press, Wallingford, Oxfordshire, England.

Simola, H. and Lodenius, M. 1982. Recent increase in mercury sedimentation in a forest lake attributable to peatland drainage. Bulletin of Environmental Contamination and Toxicology 29:298-305.

Sjora, H. 1980. Peat on earth: Multiple use or conservation? Ambio 9:303-308.

Sparling, J. 1979. Wetland panorama. Ontario Naturalist 19:10-17.

Stewart, A.J.A. and Lance, A.N. 1983. Moor-drainage: A review of impacts on land use. Journal of Environmental Management 17: 81-99.

Tallis, J.H. 1983. Changes in wetland communities. pp. 311-344. In Gore, A.J.P., ed. Ecosystems of the World, 4A Mires: Swamp, Bog, Fen and Moor. Elsevier Scientific Publishing Company, New York, USA.

Vitt, D. and Slack, N. 1975. An analysis of the vegetation of Sphagnum dominated Kettle-hole bogs in relation to environmental gradients. Canadian Journal of Botany 53:332-359.

Washburn and Gillis Assoc. Ltd. 1982. Survey of literature on the assessment of the pollution potential of the peat resource. National Research Council of Canada, Division of Energy Research and Development, Ottawa, Canada. Report NRCC 20755, 130 p.

Wiken, E.B., Welch, D.M., Ironside, G.R. and Taylor, D.G. 1981. The Northern Yukon, an Ecological Land Survey. Ecological Land Classification Series No. 6, Lands Directorate, Environment Canada. Ottawa, Ontario, Canada. 197 p.

THE IMPACT OF HYDROELECTRIC DEVELOPMENTS ON THE LECHWE AND ITS FEEDING GROUNDS AT KAFUE FLATS, ZAMBIA.

H. N. Chabwela
Dept. of National Parks and Wildlife Service, Private Bag 1
Chilanga, Zambia

G. A. Ellenbroek
Dept. of Plant Ecology, University of Utrecht
Lange Nieuwstraat 106, 3512 PN
Utrecht, The Netherlands

ABSTRACT

Dam building on the Kafue Flats ecosystem in Zambia imposes serious threats to the Kafue Lechwe. Although our knowledge of the complex wetland ecosystem of the Kafue Flats is far from complete, there is reason enough to believe that the future of the Flats and its natural inhabitants is dark. If the continued existence of the large herds of Lechwe is to be preserved, the current developments in the area must be watched carefully. In this respect it is hopeful to note that the Department of National Parks and Wildlife Service of the Republic of Zambia puts much effort in continuing the Lechwe studies on the Kafue North bank. Increased anti-poaching activities in the area may also contribute to safe-guarding the future of the Lechwe population on the Flats. Post-dam studies will add valuable information to our present knowledge of African floodplain ecosystems and may help to make wiser decisions in the many future cases of utilizing the water resources in Africa.

INTRODUCTION

The Kafue Flats, located in southern Zambia, are composed of swamps, marshes and extensive floodplain grasslands, which developed in an old lake basin. The total area from Itezi-tezi upstream to Kafue Gorge downstream of the Flats is approximately 6500 km^2. The average elevation is about 1000 m. The fall of the Kafue river channel through the Flats is extremely low, amounting to only 10 m over a channel length of about 450 km. As a result, the rising water of the main river during the rainy season causes flooding of most of the Kafue Flats in April/May. During the dry season the water slowly finds its way back to the main river and in November/December the water level reaches its annual low. In the seventies, however, the construction of dams at Itezi-tezi and Kafue Gorge for the generation of hydroelectric power strongly altered the natural flooding pattern of the Kafue Flats (FAO 1968; Werger and Ellenbroek 1980; Ellenbroek 1987). The Kafue Flats are renowned for their extremely rich birdlife. With the enormous numbers of waterfowl the Flats are listed among the best stocked bird sanctuaries of the world. By far the most important larger herbivore is the Kafue Lechwe (Kobus leche kafuensis Hallenorth) (Fig. 1). This animal, one of the four congeneric species of Zambia is entirely confined to the Kafue Flats (Ansell 1964) and is particularly well adapted to the alternating wet/dry conditions of the floodplains. Today most of the Lechwe are concentrated in the two national parks on the Flats: Lochinvar and Blue Lagoon, respectively, on the South and on the North Bank of the Kafue river (Fig. 2). Before and during the construction of the hydroelectric dams a number of ecological studies were conducted on the Kafue Flats, mainly at Lochinvar (FAO 1968; Hedberg 1971; Sheppe and Osborne 1971; Sayer and Van Lavieren 1975; Rees 1978; Ellenbroek 1987). These studies were concentrated on making predictions of the consequences of hydroelectric development to the status of the Kafue Lechwe and the Kafue Flats ecosystem as a

D. F. Whigham et al. (eds.), Wetland Ecology and Management: Case Studies, 95–101.

Figure 1. Territorial male Kafue lechwe (Kobus leche kafuensis Haltenorth) at their lekking sites on Lochinvar National Park (Photo G. Ellenbroek).

whole. As most of these studies were confined to the South Bank, the Department of National Parks and Wildlife Service decided to extend base-line studies to the North Bank in 1975. The purpose of these latter studies would be to investigate the impact of the disturbed flooding pattern on the Lechwe population in order to come up with the long term mitigating measures to safe-guard the continued protection of the Kafue Lechwe. This paper discusses the results of lechwe physical condition studies and population size estimates on the Kafue North Bank until 1985.

METHODS

Sampling was based on the normal flooding pattern with rising water levels from December to March, receding floods from April to July and low water levels during the period from August to November. The feeding grounds of the lechwe on the Kafue Flats change with the season as the large herds follow the annual movements of the water level. During the course of a year lechwe settle, for different length of time, in all of the five major vegetation zones on the Flats. These vegetation zones are described below.

Vegetation.

As mentioned above the vegetation of the Kafue Flats shows a clear zonation (Fig. 3) which is particularly well preserved in the two national parks in the area (Douthwaite and Van Lavieren 1977; Werger and Ellenbroek 1980; Ellenbroek 1987). In the study area on the North bank of the Kafue from the river to the woodland, the following vegetation zones are distinguished.

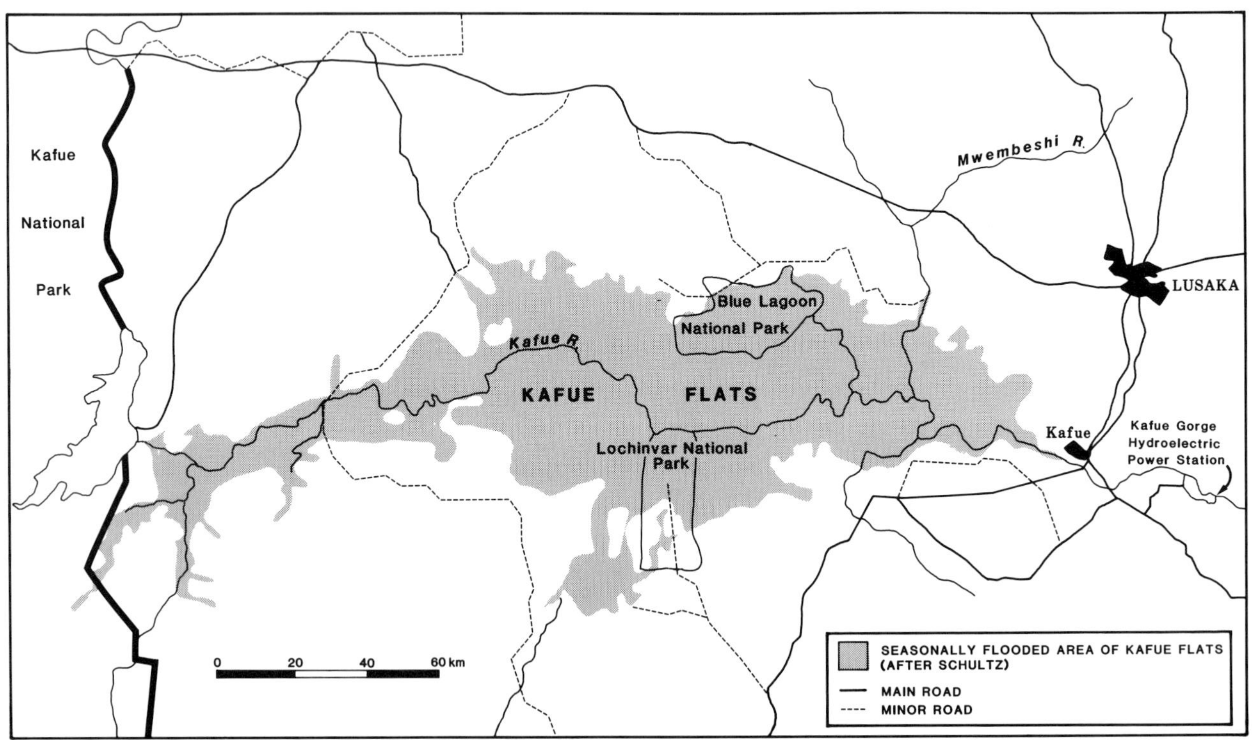

Figure 2. Situation map of the Kafue Flats and the lower Kafue Basin.

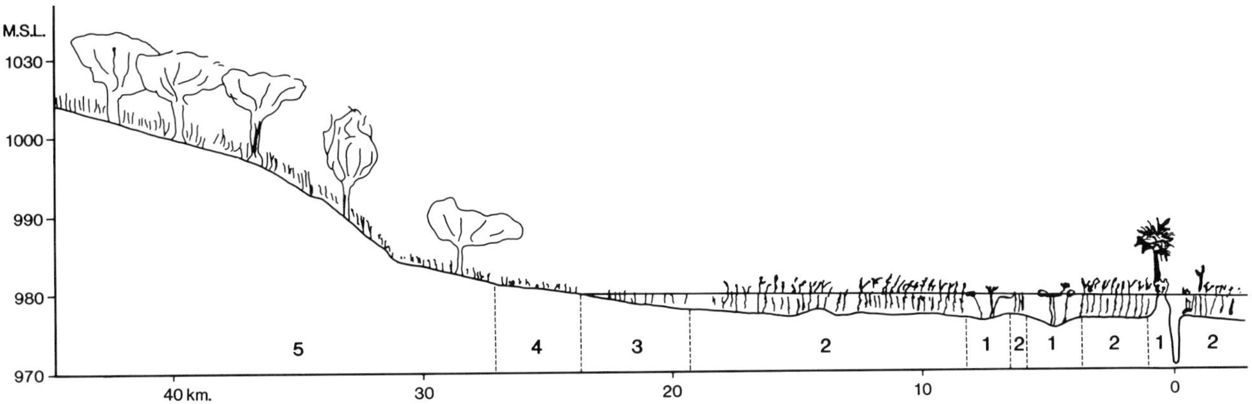

Figure 3. Schematic cross section of the Kafue river North Bank at Mutchabi, showing the major vegetation zones: 1. Levees and lagoons; 2. Tall grass floodplain grassland; 3. Water meadows; 4. Termitaria grasslands and 5. Woodland.

1. The levees and lagoons along the main river: Levees are topographically higher areas which are not flooded during hydrologically normal years. Although the levees are dominated by tall grasses such as Echinochloa sp., Vossia cuspidata Griff. and, in the vicinity of fishermen villages, Sorghum verticilliflorum (Steud.) Stapf, some scattered trees of the species Acacia albida Del. and the palm Borassus aethiopum Mart. occur too. The lagoons are mainly found along the riverine meander belt. They are usually permanently inundated, abandoned river courses with an open connection to the main river. The vegetation of the lagoons varies from mats of floating species such as Aeschynomene fluitans Peter, rooted waterlilies, communities with Nymphaea lotus L. and Nymphaea caerulea Savigny to submerged vegetation with Utricularia foliosa L. The lagoons are often fringed with tall grasses such as Vossia cuspidata, Echinochloa sp. and Oryza longistaminata Chev. & Roehr.

2. Floodplain grasslands: The Floodplain grasslands occupy the major part of the floodplain and consist of tall grasses predominantly of the species Vossia cuspidata, Echinochloa scabra (Lam.) Roem & Schult. and Oryza longistaminata (Fig. 4). Locally, Leersia hexandra Sw. and Brachiaria rugulosa Stapf may also be important. These grasslands are normally deeply flooded (up to 2.5 m) at the height of the floods. The annual net primary production of the tall floodplain grasslands is extremely high, amounting to about 4 kg m^{-2} (Ellenbroek 1987).

3. Watermeadows: The watermeadows occur on the edges of the floodplain where they are inundated only for a short period during high water levels. The main species constituting the watermeadows are short, small leaved grasses such as Panicum repens L., Leersia denudata Launert, Acroceras macrum Stapf and Paspalidium obtusifolium (Delile) N. D. Simpson. Locally, Oryza longistaminata may also be abundant.

Figure 4. Tall, deeply inundated floodplain grassland near the Kafue river at Lochinvar National Park, with Oryza longistaminata Chev. & Roehr., Vossia cuspidata Griff. and Echinochloa spp.

4. Termitaria grasslands: Termitaria grasslands occupy the area immediately above the high flood line. They are scattered with termite mounds. The area is usually only inundated for

short periods of time after heavy, local rainfall, but in years of extremely high water levels it may become flooded with river water during the early dry season. The main species of the termitaria grasslands include the grasses Setaria sphacelata (Schumach.) Stapf et Hubbard, Panicum sp. aff. porphyrrhizos Steud., Echinochloa pyramidalis (Lam.) Hitch & Chase, Eragrostis inamoena K. Schum. and Sporobolus pyramidalis P. Beauv. and a variety of herbs.

5. Woodlands: The woodlands surrounding the Kafue Flats are made up of several distinct vegetation types. The woodlands bordering the termitaria grasslands are mainly munga woodlands. These are open, park-like, one-storeyed woodlands dominated by trees such as Albizia harveyi Fourn., Acacia polyacantha Willd., Acacia galpinii Burtt Davy, Combretum spp. and with an often tall grass cover of species of the genera Hyparrhenia, Setaria and Panicum. These woodlands may become inundated for a short period during the height of the rainy season as the heavy clay soils are poorly drained.

Lechwe physical condition estimates.

Lechwe physical condition was determined using (a) Kidney fat index (Smith 1970; Sinclair and Duncan 1972); (b) Deposition of subcutaneous fat (Smith 1970) and (c) Quality of bone marrow (Sinclair and Duncan 1972). All sampled animals were examined for three indices immediately after they were shot. The kidney fat index was analyzed using t-test for the difference between two means.

Lechwe distribution and population estimates.

Aerial surveys were conducted over twelve months to map the distributions of the lechwe. Transects were flown at 100 m in a North-South direction at right angles to the Kafue river and at intervals of 4 km apart. Animal locations were plotted on a 1:250,000 map and numbers were estimated. Ground surveys by traveling either in canoe, airboat, motorvehicle or on foot were also conducted to map lechwe movements.

RESULTS

Table 1 shows the monthly distribution patterns of the lechwe before hydroelectric dam development. During the period of high water

Table 1. Monthly distribution patterns on the Kafue Fltas of lechwe before hydroelectric dam development.

Period	Levees & Lagoons	Tallgrass Floodplain Grassland	Shortgrass Floodplain Grassland	Termitaria Grassland	Woodland
December	x	x			
January		x	x	x	
February		x	x	x	
March			x	x	
April				x	x
May				x	x
June		x	x		
July		x	x		
August		x	x		
September	x	x	x		
October	x	x			

Figure 5. Lechwe herds grazing the dry Paspalidium watermeadows on the edges of the floodplain during the early rainy season in December (Photo G. Ellenbroek).

levels, lechwe were mainly found in termitaria and woodland zones. When the floodwaters receded, lechwe moved to the floodplain grazing the meadows from about June to August and the floodplain grasslands from September to December (Fig. 5). Only during low water lechwe were found on the levees and along the lagoons. After the completion of the dams, lechwe distribution patterns have considerably changed. The lechwe now spend most of their time along the lagoons and on the levees along the main river throughout the year.

Lechwe numbers have also changed markedly since the dams became operative. Between 1970 and 1976, before the construction of the dams, the size of the lechwe population on the Kafue Flats was estimated every year. The total counts varied between 47,000 and 60,000, averaging about 53,000 animals. Recent estimates, however, show a considerable decline of the lechwe population to about 40,000 in 1983 and 32,000 in 1984 (Howard 1984; Chabwela 1976; 1984).

Physical conditions estimates gave the following results. The mean kidney fat index during the period of low water amounted to 15.1 ± 7.2. Examination of the adult bone marrow showed the quality ranging from brown loose to red running marrow. There was no subcutaneous fat in all carcases examined. Comparison of these results with those of earlier studies shows that the lechwe physical conditions during the period of low water has not changed after the construction of the hydroelectric dams.

DISCUSSION

As stated above, the population size of the lechwe on the Kafue Flats has strongly fluctuated during the past decades. Detailed estimates on the size of the lechwe population are available since 1952. Lechwe were small in numbers, less than 10,000, until towards the beginning of the seventies, when numbers built up to twice the present population size. There is, however, sufficient reason to believe that factors responsible for the relatively low numbers of lechwe before the 1970s were not the same as those depressing the population size now. Darling (1960) has discussed the mass slaughter of lechwe by the local inhabitants while exercising their traditional hunting rights. This practice was banned by strict game laws in 1955 and later in 1968 which resulted in rapid growth of the lechwe population. Since the game laws have continued to be very strict, most probably the present decline of the population of the Kafue lechwe can be attributed to presence of the dams.

The results of this study support the original view that the distribution of the lechwe would be affected by the hydroelectric developments on the Kafue Flats. It is generally predicted that flood levels would be higher and flooding would be prolonged in very wet years because of the effect of the lower dam, while in the dry years floods would be lower and for a shorter period of time because of the effect of the upper dam (FAO 1968; Williams 1977; DHV 1980). The continued drought of the past three years confirms this latter view. The impact of the changed flooding regime on the ecology of the Flats and particularly on that of the Kafue lechwe is discussed below.

The relationship between the Kafue lechwe and the vegetation of the Kafue Flats before the impoundment of the dams has been studied by several researchers from different points of view (Handlos et al. 1976; Schuster 1977; Rees 1978; Ellenbroek 1987; Ellenbroek and Werger 1988). It has been argued that there exists a very strong relationship between the lechwe grazing regime, the annual flooding cycle and the development of the plant communities on the floodplain and adjacent areas (Ellenbroek 1987). The Kafue Flats are regarded as an ancient ecosystem and the Kafue lechwe, which is endemic to the Flats, has been by far the most important herbivorous species of the floodplain grasslands for a very long period of time. Therefore, it has been suggested that co-evolution of plant forms and lechwe grazing strongly determined the structure of the floodplain grasslands (Ellenbroek and Werger 1988). To understand the impact of a changed water regime on this complex relationship between grasslands and grazer, the effect of lechwe grazing on the vegetation and the role of the flooding in this process will be briefly outlined below.

The most important feeding grounds of the lechwe before the construction of the dams were the tall floodplain grasslands and the water meadows. Vegetation development in these areas strongly depended on the presence of large herds of lechwe. The deeply flooded tall grasslands were protected from grazing for a few weeks to usually several months during the period of high water and developed a lush canopy which highly emerged above the water surface (Ellenbroek 1987). The shallow flooded water meadows are, even when flooded, not or only for a very short period of time, protected from defoliation. Huge concentrations of lechwe consume almost all of the emergent plant material in these meadows and hence, most of the green matter, including the leaves, occurs submerged.

It has been stated above, that the tall floodplain grasslands are highly productive and form the most important feeding grounds for the lechwe. These grasslands, however, require at least a short period of deep flooding (50 cm or more) to reach proper development and high production rates. Continuous grazing on the floodplain will strongly alter the structure of the perennial grasscover, resulting in a great loss of productivity and decreased carrying capacity.

In the case of the water meadows, the effect of the water regulation on the Kafue Flats is less clear, the soils of the water meadows, however, hardly differ from those of the adjacent termitaria grasslands and it seems likely that when the hydrological differences which now determine the divergent plant communities disappear, the water meadows will become gradually replaced by dry termitaria grassland. Although this does not mean a loss in net primary production it does mean a loss of palatable green plant material in an area which is of vital importance to the lechwe during the early dry season in June/July.

Other negative effects of the water works on the ecology of the Kafue lechwe have been mentioned. Schuster (1977) in his behavioral studies of the lechwe showed complicated social organization of the lechwe population, closely associated with the natural flooding regime of the Flats. During the period of mating, the lechwe show the typical territorial behavior known as lekking (Fig. 1). Unlike other African antelope species the leks of the Kafue lechwe are not permanent. In the case of the Kafue lechwe the use of leks coincides with the period of rising water. The lechwe's adaptation to this is to set up a succession of temporary leks, which are abandoned when the flood rises, only to set up new ones elsewhere as the herds migrate to the higher ground. Male herds move in advance of the females and set up their leks before the females move through the area. The pattern of the distribution and succession of the leks appears to be very similar from year to year. Schuster states that breeding success is highly dependent on the leks as mating outside of leks only seldom occurs. He argues that changes in the natural flooding regime will interfere with the closely related lekking behavior of the Kafue lechwe and that the high breeding rate, which is necessary to sustain the large herds, will probably not be maintained.

Another plausible explanation for the decreasing lechwe numbers on the Kafue North bank may be found in the increased accessibility of the floodplain. Illegal hunting, directly affecting the numbers of lechwe, and increased fishing activities, causing continuous disturbance of the lechwe herds particularly during the period of lekking, might well be responsible for a serious population decline.

ACKNOWLEDGEMENTS

The authors wish to thank the biologists D. Chimbali, P. Sichone, the warden M. Malama and the senior ranger, late F. Chitukole, for assisting in collecting the data. We sincerely express our gratitude that the Ministry of Lands and Natural Resources through the Department of National Parks and Wildlife Service and the Kafue Basin Research Committee of the University of Zambia have continued to support the Kafue Flats research.

REFERENCES

Ansell, W. H. F. 1964. The Kafue lechwe, Puku 2: 10-13.

Chabwela, H. N. 1976. Research and management. Annual report. Department of National Parks and Wildlife Service. Chilanga, Zambia. 41 p.

Chabwela, H. N. 1984. Ecological adaptation of four sympatric species of Kobus as related to human induced disturbances. Survey data. Chilanga, Zambia.

Darling. F. 1960. Wildlife in African territory. Oxford University Press, London, UK. 160 p.

D. H. V. 1980. Kafue Flats hydrological studies, Final Report. DHV Consulting Engineers. Amersfoort. The Netherlands. 198 p.

Douthwaite, R. J. and Van Lavieren, L. P. 1977. The vegetation of Lochinvar National Park. National Council for Scientific Research. Lusaka, Zambia. 66 p.

Ellenbroek, G. A. 1987. Ecology and productivity of an African wetland system. The Kafue Flats, Zambia. Dr. W. Junk

Publishers. Dordrecht, The Netherlands. 267p.

Ellenbroek, G.A. and Werger, M.J.A. 1988. Grazing, Canopy structure and production of floodplain grasslands at Kafue Flats, Zambia. pp. 331-337. In Werger, M.J.A., van der Aart, P.J.M., During, H.J., and Verhoeven, J.T. A., eds. Plant form and vegetation structure. SPR Academic Press. The Hague, The Netherlands.

F.A.O. 1968. Multipurpose Survey of the Kafue River Basin, Zambia. Vol. I-V. U.N.F.A.O., Rome, Italy. 1024 p.

Handlos, D. M., Handlas, W. L. and Howard, G. J. 1976. A study of the diet of the Kafue lechwe (Kobus lechwe) by analysis of rumen contents. Proceedings 4th Regional Wildlife Conference of East and Central Africa. Department of National Parks and Wildlife Service. Lusaka, Zambia. pp. 197-211.

Hedberg, O. 1971. Kafue River Hydroelectric Power development stage II. SWECO. Lusaka, Zambia.

Howard, G.W. 1984. Personal Communication. University of Zambia. Lusaka, Zambia.

Rees, W.A. 1978. The ecology of the Kafue lechwe. Journal of Applied Ecology 15:163-217.

Sayer, J.A. and Van Lavieren, L. P. 1975. The ecology of the Kafue lechwe population in Zambia before the operation of hydroelectric dams on the Kafue river. East African Wildlife Journal 13: 9-38.

Schuster, R. H. 1977. Social organization of the Kafue lechwe. Black lechwe 12, 3:40-51.

Sheppe, W. and Osborne, T. 1971. Patterns of use of a floodplain by Zambian mammals. Ecological Monographs 41: 179-205.

Sinclair, A.R. E. and Duncan, P. 1972. Indices of condition in tropical ruminants. East African Wildlife Journal 10: 143-149.

Smith, N. S. 1970. Appraisal of condition estimation methods for East African Ungulates. East African Wildlife Journal 8:123-129.

Werger, M. J. A. and Ellenbroek, G. A. 1980. Water resource management and floodplain ecology: An example from Zambia. pp. 693-702. In Furtado, J. I., ed. Tropical Ecology and Development. International Society for Tropical Agriculture, Kuala Lumpur, Malaysia.

Williams, G. J. 1977. The Kafue hydroelectric schemes and its environmental setting. pp. 13-27. In Williams, G.J., and Howard, G.W., eds. Development and Ecology in the Lower Kafue Basin in the nineteen seventies. Teresianum Press, Lusaka, Zambia.

SHOULD THERE BE MAN-MADE LAKES IN AFRICA?

Dr. J. P. Msangi[1]
Department of Geography
University of Dar es Salaam, Tanzania

Dr. G. A. Ellenbroek
Department of Plant Ecology
University of Utrecht, The Netherlands

ABSTRACT

For many good reasons , a large number of man-made lakes have been constructed in Africa. Unfortunately, many of the optimistic predictions about the positive benefits of the lakes have had to be modified because of unforeseen problems. In this paper, we discuss the indirect costs associated with the lakes and their wetlands and consider a variety of ecological issues and problems (health, social changes, political differences) that have arisen.

[1] Current address:
Geography Department
Egerton University
P.O. Box 536
Njoro, Kenya

INTRODUCTION

The creation of man-made lakes behind dams constructed across river valleys has become a popular method of developing the water resources of river basins (Lowe-McConnell 1966; Lagler 1969; Obeng 1969; Ackermann et al. 1973). The purposes for their creation vary from river regulation for flood control, irrigation, navigation, fishing, hydroelectric power production, and water supply either for industrial or domestic purposes. Surveys prior to the dam construction, as well as the actual construction and post-construction implementation of infrastructures require large amounts of money. In order to justify such expenditures, several purposes are combined to give multi-purpose projects. Fishing and lake shore-settlements which in most cases have been direct consequences of the created lakes are often considered as secondary purposes, but never as primary.

This paper concentrates on man-made lakes in Africa. Conditions in Africa, though similar to other tropical or underdeveloped regions, are different from those in temperate or developed regions.

Geomorphological Aspects

Africa presents certain advantages for the construction of dams (Warren and Rubin 1968). In general, Africa is an ancient continent composed of some of the world's most ancient and impermeable rocks. Most of the continent is formed by plateaux which have well-marked edges. Rivers fall over these and such sites provide points for power dams. Erosional forces have been in action for so long that numerous breaks of slopes and gorges have been formed and these provide possible sites for dams. Although the ancient rocks may 'rot' superficially (through chemical weathering) and give rise to constructional problems, they usually provide a firm anchor for dams and provide good building material. Extensive fertile plains adjacent to these rivers have also been formed through erosion and weathering. These provide rich agricultural and grazing lands.

Most African rivers have very uneven flow regimes so that their volume and velocity

D. F. Whigham et al. (eds.), Wetland Ecology and Management: Case Studies, 103–108.

fluctuate greatly from season to season and from year to year. Due to the fact that most of the rocks are ancient and impermeable or have very low permeability, there is usually little base flow to support very high river discharges during dry seasons. Yet solar radiation is generally high throughout the year and thus encourages quick and continuous growth. This means very high evapotranspiration rates throughout the year. Therefore to guarantee an all year supply of water for any purpose and especially for combined purposes, retention of the high flows to raise river discharge during low flow periods or to meet combined demands is necessary. From this point of view man-made lakes SHOULD BE created to solve the problem of conflicting and excessively high demands.

The African man-made lakes have been constructed from the above point of view. Physically it was possible and there was a need for them.

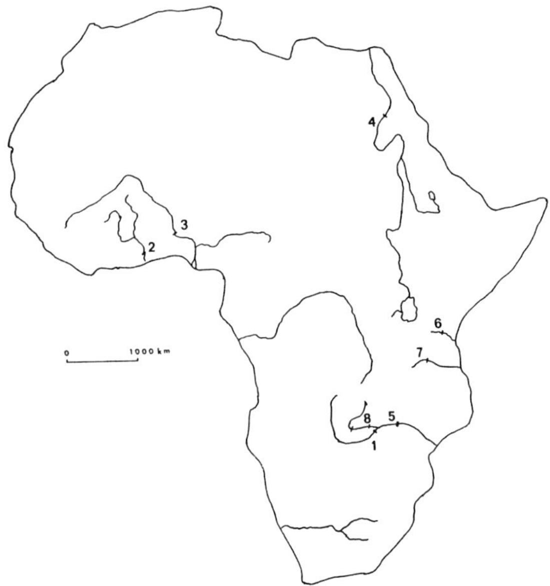

Figure 1. Location of the large man-made lakes in Africa.
1. Lake Kariba; 2. Lake Volta; 3. Lake Kainji; 4. Lake Nasser; 5. Cabora Bass Reservoir; 6. Nyumba ya Mungu Reservoir; 7. Mtera Dam; 8. Kafue Flats.

Larger Man-Made Lakes in Africa

So far there are several large man-made lakes in Africa (Fig. 1) that have been in existence since the late 1950s. These lakes have played a vital role in the process of economic development, or rather have been expected to do so. In each African country in which a major lake has been created, or for which one has been planned, the project has entailed not only unparalleled commitment of economic resources, but also important social, physical and political implications.

The largest man-made lake to be constructed in Africa was the Kariba lake behind Kariba dam built across the Zambezi River (Fig. 1). This lake, formed in 1958, and located on the border between Zambia and Zimbabwe (formerly N. and S. Rhodesia), has a volume of 160 km^3. Its main purposes were to provide hydroelectric power, a navigational routeway and fishing. It has had a considerable impact on the people and the areas surrounding it.

Volta lake at Akosombo, Ghana, created in 1965, was the second largest man-made lake in Africa. This lake has the largest surface area in the world but a capacity of only 165 km^3. Its main objective was to provide hydroelectric power for the mining and smeltering of Ghana's aluminum resources and to stimulate general industrialization of the country. Other purposes were to provide navigational routeways and to establish agricultural settlements along its shores.

The Kainji lake, located in Nigeria across the Niger River and formed in 1968, is the third largest man-made lake in Africa. It forms the first stage of a program to provide electric power for industrialization in Nigeria. Other purposes included navigation facilitation on the river, flood control, fishing and the improvement of road communication.

Lake Nasser, behind the Aswan High dam in Egypt, started filling in 1970 and is now the second largest man-made lake in the world. Its primary objective was water storage and conservation through flood control. Today the water is mainly used for irrigation and hydroelectric power generation.

The newest man-made lake in Africa and the third largest in the world is the Cabora Bassa behind Cabora Bassa dam across the Zambezi River at Kebrabassa (Middlemas 1978). Completed in 1975, the lake is designed to produce 4000 MW of power, 90% of which is for sale to South Africa. Its main purpose is to generate power for mining and general industrialization of the country. Other purposes include making the Zambezi River navigable across the continent to Angola, linking Zimbabwe, Zambia, Malawi and Mozambique.

Among the smaller man-made lakes which

dot the continent is Nyumba ya Mungu in Tanzania on the Pangani River (Denny and Bailey 1976; Kaduma 1977). The dam was completed in 1965 and the lake took two and a half years to fill. The main purposes of the lake were to regulate the river, to provide water for irrigation and hydroelectric power production. Other intended uses were fishing and tourism. Also found in Tanzania is Mtera lake whose purpose is to regulate Great Ruaha River flow for the power station located at Kidatu.

Figure 2. Itezi-tezi reservoir in Zambia two years after the filling of the lake. Drowned trees of the former woodland on the hill slopes still protrude above the water. (Photo G. Ellenbroek).

In Zambia the construction of dams in the Kafue River was completed in the 1970s. A dam in the mid-Kafue at Itezi-tezi (Fig. 2) and a second dam in the lower Kafue River at Kafue Gorge allow water regulation of the large floodplain areas of the Kafue Flats (Werger and Ellenbroek 1980; Ellenbroek 1987). The water is almost exclusively used for hydroelectric power generation through a power plant downstream of the Gorge dam. The electric power is mainly used for the Zambia's copper mining industry in the upper part of the Kafue Basin.

Figure 3. Changes in the natural flooding regime following hydroelectric development in the Kafue river in Zambia forced fishermen to abandon their villages on the flooded levees (Photo G. Ellenbroek).

Costs of Economic Progress

As stated before, man-made lakes have solved critical problems caused by uneven regimes like flooding. They have conserved the wet period flows for use during critical dry periods. They have provided a much needed protein in some parts of Africa like Ghana where cattle rearing is limited by tsetse flies. They have provided great amounts of cheap hydroelectric power. <u>But</u> one question we must ask ourselves is "<u>at what cost</u>?". Not only costs for dam construction and for establishing the other facilities for power, fishing, irrigation and water supply. Here we mean the secondary and indirect costs.

Mostly these costs are what the environment has to pay in order to achieve 'economic progress'. These costs are such as the lost land which is covered by the lake (Fig. 3), disrupted way of life of the displaced population, introduction of water borne and water related diseases, or accelerated rate of increase of these diseases (due to the creation of ideal environmental conditions by the lake) and disruption of established ecosystems.

Ecological Effects

Damming and water regulation greatly disturb the natural equilibrium in a river ecosystem. This has resulted in many ecological problems which have even affected the desired economic progress. Upstream and impoundment flow velocities are reduced and silt, carried down with the river water is trapped. This has caused not only damage to agricultural land use but to the fishing industry as well.

In many cases, bush clearing of areas bound to become submerged after closure of a dam has not been carried out. The result of such negligence is that the decaying, drowned vegetation mass on the bottom of the reservoir uses up all the available dissolved oxygen and creates anaerobic conditions which are strongly deleterious to fish.

Invasion of aquatic weeds in man-made lakes in Africa has caused problems in many ways. Newly filled lakes have been subjected to explosive growth of surface floating species such as water fern (<u>Salvinia molesta</u>), water hyacinth (<u>Eichhornia crassipes</u>) and water lettuce (<u>Pistia</u>

stratiotes). Particularly the alien species Eichhornia and Salvinia have become a nuisance and a serious threat to the proper functioning of water works in impoundment areas. Blocking spillways, interfering with river flows, navigation, stock drinking and in the case of Eichhornia, causing strongly increased water losses through excessive evapotranspiration are wide-spread problems reported from all parts of Africa (Mitchell 1975; 1985). The development of submerged species and emergent macrophytes is greatly inhibited through the shading of the water surface by the tightly packed floating mats.

Damming, however, does not only affect the area upstream of an impoundment, it also interferes with the riverine ecosystem downstream. As stated before, most of the impoundment areas in Africa are situated in areas with a strong seasonality in rainfall distribution. Thus the affected rivers often show a strong seasonal variation in water discharge. Under natural conditions peak water levels in these rivers lead to the flooding of large areas of adjacent land (see above). These seasonally flooded areas, often consisting of highly productive floodplain grasses such as Vossia cuspidata, Echinochloa spp. and Oryza spp., are dependent on an annual period of inundation (Thompson 1985; Ellenbroek 1987). They support a rich fish population and large numbers of birds and game animals. Moreover, floodplain grasslands are often of invaluable importance to dry season grazing of domestic stock. Water regulation, largely preventing flooding, causes the destruction of these grasslands, resulting in a decrease in the carrying capacity of the area and its dependent herbivores. Deplorable examples of the deterioration of such floodplain ecosystems after artificial changes in the water regime are known from many parts of Africa (*cf.* Attwell 1970; Werger and Ellenbroek 1980; Denny 1985).

Health Problems

Apart from flooding rich grazing and/or agricultural lands and sometimes valuable forests, the shorelines of these lakes have become the home of water-borne disease vectors (Stanley and Alpers 1975). The land adjacent to the shoreline of the lakes provides ideal environmental conditions for *schistosomiasis* and *opisthorchiasis* snails, *trypanosomiasis* flies (tsetse), malaria, *filariasis*, yellow fever, *encephalitis* and dengue mosquitoes. These diseases spread fast due to high human concentration along the lake shores (mostly fishing communities). The luxurious vegetation which has developed along the

Figure 4. Improved fishing conditions in impoundment areas often looked promising but in most cases the negative effects on the riverine ecosystems were strongly underestimated or not considered at all (Photo G. Ellenbroek).

shoreline due to a raised water table is ideal grazing land for larger herbivores which also encourages the development of tsetse flies. The shallow and relatively still marginal waters provide an ideal habitat for *schistosomiasis* and *opisthorchiasis* snails. Yet here is where man takes his animals for drinking, lands his catch, draws his water from, washes his clothes and takes his bath. Small children paddle here too, and housewives may wash dishes as well. Due to the poor sanitation which characterizes these fishing communities, it is in the bushy shore vegetation and sometimes in the shallow marginal waters where man relieves himself.

Social Constraints

In Africa, where the river valleys provide one of the best lands for agriculture with a guaranteed water supply for domestic uses all year round, displaced population had been high. This includes some 52,000 at Kariba, 78,205 at Volta, 120,000 Aswan, 50,000 Kainji and 24,000 at Cabora Bassa. Some of these people, for example the Nubians from Aswan, were moved from Wadi Halfa to Kashim el Girba several hundred miles upstream. At Kariba the Tonga (30,000) were moved from their rich alluvial soils to poor hilly soils. At Kainji, the people moved to higher ground can not utilize the lake water for irrigation (this being a semi-arid region of N. W. Nigeria) because of lake level fluctuations (about 9 meters annually) and a lack of installed water pumps. Even downstream areas where about 200,000 hectares were freed from flooding, no irrigation agriculture techniques have been introduced by 1978 (10 years later). Hence when rain fails these people plus their livestock are faced with famine. All these displaced people have been moved from their places of birth to remote and strange areas where they had to start a new life. They needed to adapt to the change. They abandoned their homes and farms and ancestral graves and familiar grounds to the floods and ample water (even if its only for part of the year). In the case of the Nubians (Lake Nasser) and those at Volta they found houses in the new towns but they were not homes. Farmers had to cultivate new lands and crops, while fishermen had to fish a new large lake for which their gear and methods were at first inadequate (Fig. 4). In general, it took a long time for the people to adjust to the new conditions, develop social and economic stability and reestablish community and cultural identity.

In the new areas there have been many human problems due to inadequately prepared settlement sites. The settlement sites were in remote areas, cut off from other developed areas. The houses built for the displaced people were too small in the case of Lake Volta, too poor in the case of Lake Nasser and in the case of Kariba no houses awaited the newcomers. They had to build in haste, and hence the houses were of poor quality and small in size. Public services like schools, hospitals, roads and water supply were either lacking or were inadequate. Due to this fact, mortality rates were high for the first and second year. Livestock died (for example at Aswan and Kariba) while personal belongings were damaged.

Compensation has not been given to the satisfaction of displaced persons. Officials often have failed to evaluate the personal belongings left behind. Government policies on compensation vary so that in some cases, compensation was in the form of money, as in Aswan case, while in the Kariba and Volta cases, compensation was on a material basis (i.e. house for house, land for land, etc) which owners often resented when resettlement houses turned out to be smaller and of different make and arrangement from the lost houses. In the Kariba (south bank) case, the people were left to settle down as they pleased. No compensation was given; only a piece of land, grass for thatching, moving belongings in trucks and food and tax suspension for two years.

All these problems caused so much stress and misery that people left the new settlements at Volta and Aswan. Low morale affected the people's potential productivity so that the economic transformation desired by planners was not immediately achieved. It has been disclosed by several researchers that in many cases, no real progress in the life of the rural population has been obtained by lake creation because in all cases settlement along the shores and fishing were anticipated, but no communication links were established before settlement and fishing activities began. This has been a bottleneck to progress in all cases.

Political Interests

The utilization of some man-made lakes is made difficult due to different political systems in the riparian states. Most of the African large man-made lakes are formed on international rivers. During the partition of Africa, rivers were sometimes used to mark boundaries of states, as in the case of Zambia and Zimbabwe, and sometimes one river basin was divided into several states so that today most of the African large rivers are international rivers. As such, their utilization is governed by international law as set out in Article 38 of the Statute of the International Court of Justice (i.e. international customary law, international conventions and general principles of law recognized by civilized nations). The governing principle most relevant to the problems involved in the exploitation of water common to two or more states is Sovereignty. On the level of international customary law, the principle of sovereignty remains dominant in the sense that joint efforts to develop international rivers must be based on the consent of the state within whose territorial sovereignty part of the resources fall. Unless consent is forthcoming, the state wishing to develop common resources may face problems of political differences which abound in many African states today. Agreements to create man-made lakes have been reached after very long meetings and discussions. Even after creation, utilization is made complex by these differences. Lake Kariba, and Lake C. Bassa provide the best examples. In early days, political hostilities placed transportation, power production and utilization on very difficult conditions, especially so at Cabora Bassa lake where power utilization depended on Mozambique's, Zimbabwe's and South Africa's political stability. Political instability in Zimbabwe made Zambia so uncertain about continued power supply from the south bank station that she constructed other man-made lakes on the Kafue River. This has proven to be a duplicated effort and unnecessary expenditure as power has continued to flow and a second power station on the northern bank of Kariba lake has been constructed. The same problem led to the closure of the borders between Mozambique and Zimbabwe so that although the Mozambique electricity grid was extended to the border, up to now there are no power sales to Zimbabwe. The point of making the Zambesi River navigable across the continent to Angola remains a dream due to differing political stands between Mozambique, Zambia, Angola and Zimbabwe.

CONCLUSION

Various suggestions on solving the above problems have been made. Suggestions include carrying out multidisciplinary research to cover various aspects like soils, agriculture, flooding, bush clearing, aquatic weeds and their control, the people to be displaced, the new areas of resettlement and river hydrology. This research is recommended to cover a substantial zone of land around the lake which will come under its influence. This zone, like the lake itself, needs advance planning as well as post impoundment management. What all this research means is added costs for the same benefits.

Other suggestions include making the lakes environmentally safe (Obeng 1977). Such efforts would mean preventing contact of the shore population with snail infested areas by providing alternative sources of water supply, preventing lake contamination with human waste through better sanitation and waste disposal systems, preventing weed growth on lake shores, preventing upstream pollutants from reaching the lake and extending health education to the fishing communities. Again this means added costs for the same benefits. Therefore, should man learn to adapt to natural conditions like uneven river regimes and dry periods or should man try and tame these rivers and create man-made lakes? Should man look for alternative energy sources like solar radiation, wind-power, bio-gas etc. or continue to create the above problems and spend his energy trying to solve them? Is the said cheap hydroelectric power so cheap after all? Is flood control and claimed irrigation agriculture so vital? (Especially bearing in mind that NONE of the planned irrigation agriculture to produce the much needed extra food to feed the expanding populations has been implemented except in Egypt.) Hence the multipurpose aspect of these projects is not really true. The main reason is cheap power. But power for whose benefit: the local communities, the regions where projects are located, the nation or multi-national undertakings extended to Africa?

REFERENCES

Ackermann, W. C., White, G. F., and Worthington, E. B. (Ed.). 1973. Man-made lakes: Their problems and environmental effects. American Geographical Union. Washington, D. C., USA. 487 p.

Attwell, R. I. G. 1970. Some effects of Lake Kariba on the ecology of a floodplain of the Mid-Zambezi Valley of Rhodesia. Biological Conservation 2, 3: 189-196.

Denny, P. 1985. The ecology and management of African wetland vegetation. Junk. Dordrecht, The Netherlands. 344 p.

Denny, P. and Bailey, R. G. (Eds.). 1976. A biological survey of Nyumba Ya Mungu Reservoir, Tanzania. 253 p.

Ellenbroek, G. A. 1987. The ecology and productivity of an African wetland system: The Kafue Flats, Zambia. Junk. Dordrecht, The Netherlands. 267 p.

Kaduma, J. D. 1977. Man-made lakes: Their social economic and ecological impacts; The case of Tanzania. Ph.D. Thesis, University of Dar-es-Salaam, Tanzania.

Lagler, F. K. (Ed.). 1969. Man-made lakes: Planning and development. F. A. O. Rome, Italy. 71 p.

Lowe-McConnell, K. 1973. Man-made lakes. Academic Press, London, U.K.

Middlemas, K. 1973. Cabora Bassa: Engineering and politics in Southern Africa. Weiden Feld and Nicholson.

Mitchell, D. S. 1973. Aquatic weeds in man-made lakes. pp. 606-611. In W. C. Ackerman, G. F., White, and E. B., Worthington, eds. Man-made lakes: Their problems and environmental effects. American Geographical Union. Washington, D. C., USA.

Mitchell, D. S. 1985. African aquatic weeds and their management. pp. 177-202. In P. Denny, ed. The ecology and management, of African Wetland vegetation. Junk. Dordrecht, The Netherlands.

Obeng, L. E. (Ed.). 1969. Man-made lakes. Ghana University Press. Accra, Ghana.

Obeng, L. E. 1977. Should dams be built? AMBIO VI, 1.

Stanley, N. F. and Alpars, M. P. (Eds.). 1975. Man-made lakes and human health. Academic Press. London, U.K.

Thompson, K. 1905. Emergent plants of permanent and seasonality flooded wetlands. pp. 43-107. In Denny, P., ed. The ecology and management of African wetland vegetation. Junk. Dordrecht, The Netherlands.

Warren, W. M., and Nubin, N. 1968. Dams in Africa. Frank Cass & Co. Ltd.

Werger, M. J. A., and Ellenbroek, G. A. 1980. Water resource management and floodplain ecology: An example from Zambia. pp. 693-702. In Furtado, J. I., ed. Tropical ecology and Development. International Society for Tropical Ecology. Kuala Lumpur, Malaysia.

SURVEY OF MACROPHYTIC VEGETATION IN NORTH GERMAN WATER COURSES

W. Herr
D. Todeskino

IBL·Institut für Angewandte
Biologie Landschaftsökologie
Unterm Berg 39
D-2900 Oldenburg, FRG

G. Wiegleb

Universität Oldenburg
Fachbereich Biologie
Postfach 25 03
D-2900 Oldenburg, FRG

ABSTRACT

Results of a survey of macrophytes in North German riverine wetlands are presented. Species distribution is related to geographical factors and zonation patterns along a water course are discussed. Riverine wetlands are classified into community types which are then discussed from a theoretical point of view. Consideration is given to wetland conservation based on the "regionally differentiated species deficit model" that is based on "potential natural vegetation." Portions of many rivers are in a highly degraded condition and there are immediate needs for protection and restoration. Finally, the legal methods that could be used for nature conservation are discussed.

INTRODUCTION

Lower Saxony is the second largest (45,000 km^2) state in the Federal Republic of Germany. There are about 30,000 km of water courses, one third of which belongs to the macrophyte region (Roll 1938). The vegetation in running waters of the state exhibits a great diversity. We conducted a survey of this vegetation between 1978 and 1982. Results of the survey are summarized in this paper and in Herr et al. (1989), Herr and Wiegleb (1985), Wiegleb and Herr (1983, 1984b). Geographical areas are shown in Fig. 1 and the main riverine drainage systems in Fig. 2.

The objectives of the survey were:

1. Conduct a floristic inventory of species in the region and map their distribution.
2. Develop a classification of the vegetation and compile information on the distribution and ecology of each vegetation type.
3. Conduct an evaluation of the water courses with respect to nature conservation in the different geographical areas and main drainage basins in Lower Saxony.

METHODS

There are no proven methods for surveying vegetation in running water (Herr 1984). The choice of methods is almost always a compromise based on the different aims of a study. For most purposes, cover is more reliable than frequency, as frequency seems to be more subjectively defined, while cover gives a clearer idea of the real vegetation picture. On the other hand, cover can only be estimated accurately in short river sections, while frequency is more reliable in large sections. In this study, we divided the rivers into sections with a minimum length of 50 -100 m (Wiegleb 1983a).

Vegetation was sampled in 1,043 river sections. These sections were irregularly distributed along the water courses. All submerged and floating species were identified and beginning in 1980 we estimated cover of each species using the decimal scale of Londo (1984). This scale is quasi-metric and suitable for computations. Data from 204 relevees of other authors (mainly Grube 1975; Weber 1976; Weber-Oldecop 1970, 1971). were also used.

D. F. Whigham et al. (eds.), Wetland Ecology and Management: Case Studies, 109–116.

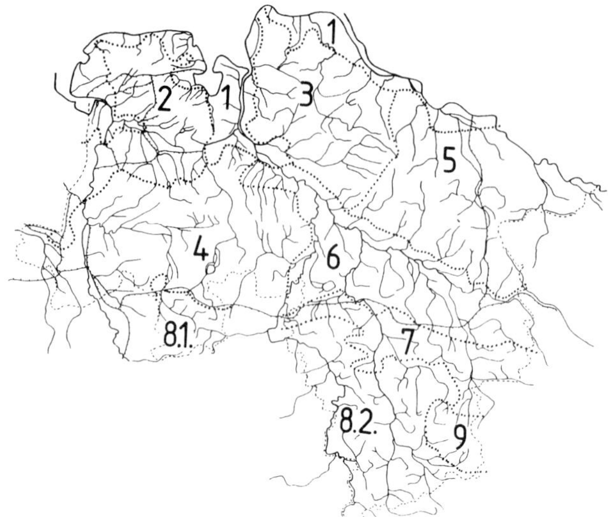

Figure 1. Main geographical areas of Lower Saxony. The northwest German lowlands are divided into:

1. Coastal marshes,
2. Oldenburg-East Frisian diluvial plain,
3. Stade diluvial plain,
4. Ems-Hunte diluvial plain,
5. Lueneburger Heide,
6. Weser-Aller lowlands.

The area of the highlands, which essentially means the area south of the Mittelland Canal is divided into:

7. Boerde (loesses) area,
8.1. Osnabrueck hill country,
8.2. Weser-Leine highlands,
9. Harz Mountains.

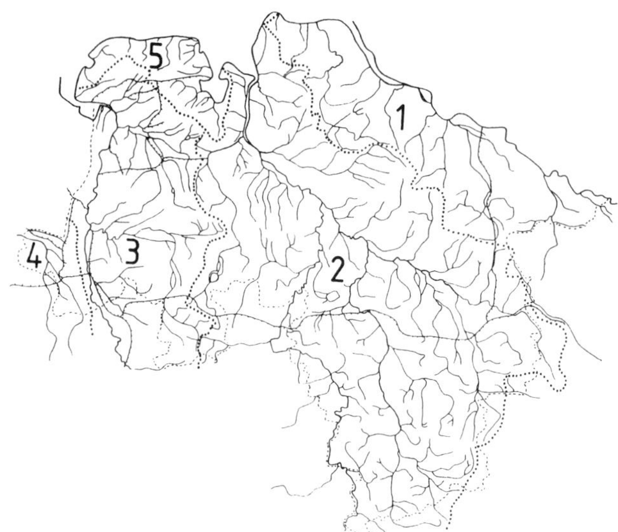

Figure 2. Main river systems of Lower Saxony

1. Elbe area,
2. Weser area, including the large subarea of Aller and Leine,
3. Ems area,
4. Vechte area,
5. Coastal areas.

Species distribution maps were compiled. The relevé data were analyzed to identify vegetation types and to carry out different numeric procedures concerning species distributions. Vegetation types were defined by growth forms of the macrophytes and displayed on distribution maps.

RESULTS

Mapping of Species

In Figs. 3-8 the author's findings are indicated by black circles, the findings of other authors by open squares. The 90 hydrophyte species can be categorized into six types based on distribution:

1. Species that occur all over the state and are very common, based on frequency of occurrence. The most common species are Callitriche platycarpa, Potamogeton pectinatus, P. crispus, and Berula erecta (Fig. 3).

2. Species that are primarily restricted to and common in lowlands. Sparganium emersum (Fig. 4), Lemna minor, Ranunculus peltatus and Elodea canadensis are common in the more continental eastern part of the lowlands. Nuphar lutea, Potamogeton natans, P. pusillus and Sagittaria sagittifolia show a more western (oceanic) distribution.

3. Lowland species that occur less frequently and have interesting distribution patterns. Four subgroups can be distinguished:

a) Species that only occur in rivers with slow current velocities (e.g. Lemna gibba, L. trisulca, Ceratophyllum demersum (Fig. 5), Spirodela polyrhiza, Polygonum amphibium, Potamogeton trichoides). Some species in this subgroup are restricted to marshes and bogs (e.g. Potamogeton compressus, Hydrocharis morsus-ranae, Utricularia australis, Myriophyllum verticillatum). b) Species whose distribution has been reduced because of pollution (e.g. Potamogeton lucens and P. perfoliatus). c) Species that only occur in rhitral water sections (e.g. Myriophyllum alterniflorum (Fig. 6), Ranunculus penicillatus, and Leptodictyum

Figure 3. Distribution of Berula erecta (Huds.) Coville in Lower Saxonian water courses. Solid circles are data from the study while open squares are data from authors cited in the text.

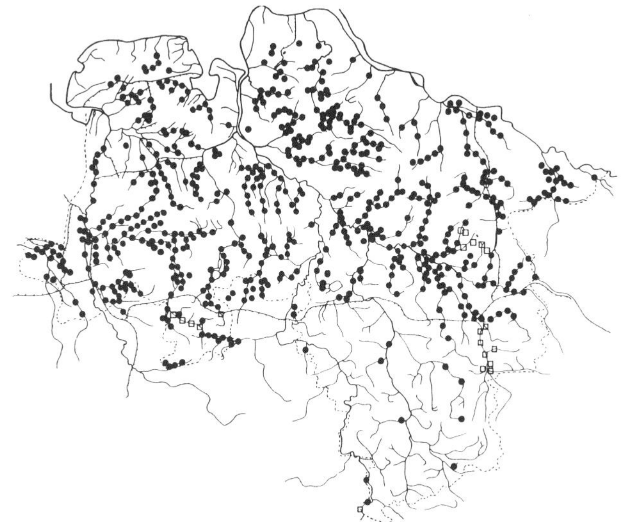

Figure 4. Distribution of Sparganium emersum Rehm. in Lower Saxonian water courses. Solid circles are data from the study while open squares are data from authors cited in the text.

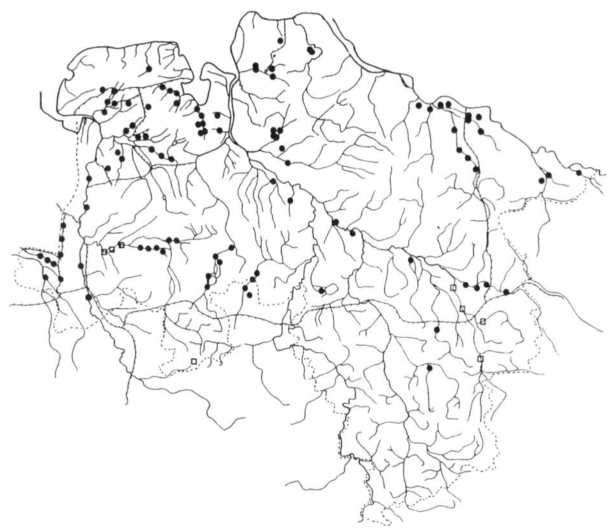

Figure 5. Distribution of Ceratophyllum demersum L. in Lower Saxonian water courses. Solid circles are data from the study while open squares are data from authors cited in the text.

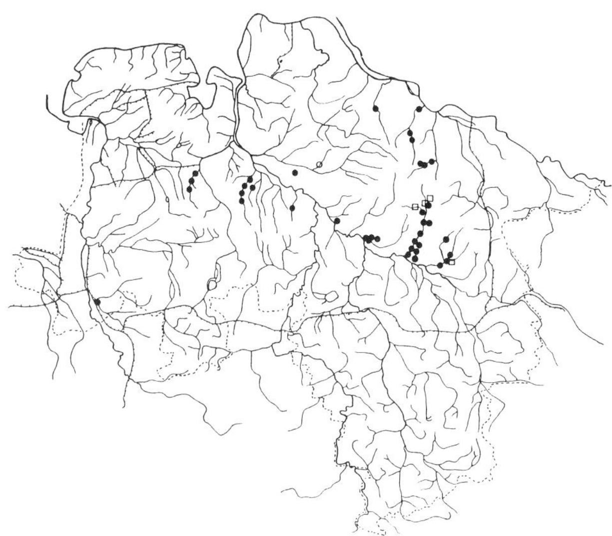

Figure 6. Distribution of Myriophyllum alterniflorum DC. in Lower Saxonian water courses. Solid circles are data from the study while open squares are data from authors cited in the text.

riparium). Several of these species have an oceanic distribution (e.g. Potamogeton alpinus and Nitella flexilis). d) Invading species, like Elodea nuttallii and Callitriche obtusangula, which come from the oceanic regions of southwest Europe. Elodea nutallii (Fig. 7) is closely related to group 3a. C. obtusangula shows a more rhitralic pattern and seems to replace C. hamulata in some river systems. 4. Bipolar species distributed in the highlands and the rhitral water courses of the lowlands but not in the Börde area. Examples are Callitriche hamulata (Fig. 8), Ranunculus fluitans and Fontinalis antipyretica. 5. Highland species

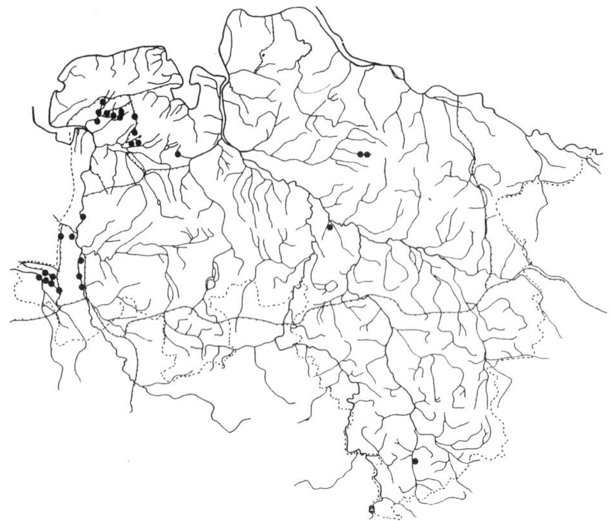

Figure 7. Distribution of Elodea nuttallii (Planch.) St. John in Lower Saxonian water courses. Solid circles are data from the study while open squares are data from authors cited in the text.

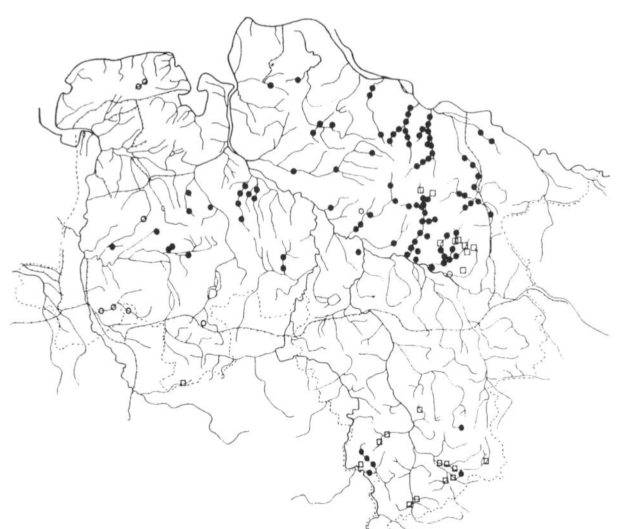

Figure 8. Distribution of Callitriche hamulata Kuetz. in Lower Saxonian water courses. Solid circles are data from the study while open squares are data from authors cited in the text.

with isolated locations in the lowlands. Only Zannichellia palustris and Ranunculus trichophyllus belong to this group. 6. Species with scattered distribution (e.g. Myriophyllum spicatum, Butomus umbellatus, Callitriche stagnalis, and Potamogeton friesii).

Figure 9. Distribution of nymphaeid-rich vegetation type in Lower Saxonian water courses. Solid circles are data from the study while open squares are data from authors cited in the text.

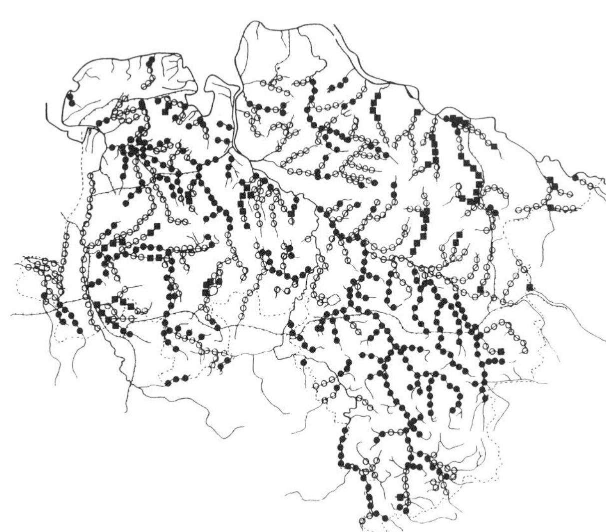

Figure 10. Assessment of Lower Saxonian water courses for nature conservancy purposes. Squares: Actual vegetation of high value for protection. Blank circles: Actual vegetation of minor value for protection. Solid circles: Actual vegetation sparse or absent.

Species that occurred in less than 10 releves were not listed in the six categories. Additional distribution maps are given in Wiegleb and Herr (1983), Wiegleb and Herr (1984b) and Herr and Wiegleb (1985).

CLASSIFICATION

Vegetation types as defined in this paper are not synonymous with associations of the Braun-Blanquet system. Two kinds of vegetation types were distinguished:

1. Comprehensive types, defined by structural and dynamic criteria (Wiegleb 1981).

2. Small types, defined by floristic criteria such as dominance and frequency (Herr 1984, Wiegleb 1983b).

Neither is this system a hierarchical one as a small type may be joined to different comprehensive types by ecological and dynamic criteria. Nine comprehensive types were recognized based on the dominant genera or species:

1. nymphaeid-rich, 2. batrachid-rich, 3. Ranunculus fluitans-rich, 4. magnopotamid-rich, 5. Potamogeton pectinatus-rich, 6. pleustophyte-rich, 7. Zannichellia- Groenlandia densa, 8. Littorelletea, and 9. Elodea-rich.

The criteria used to define each comprehensive type are given in Herr et al. (1987) and an example is given here for the nymphaeid-rich type. One of the four species (Sparganium emersum, Potamogeton natans, Nuphar lutea and Sagittaria sagittifolia) covers more than 3 on the Londo scale and one or more of these four species cover more than all other species, respectively. This type is characteristic of the North German lowlands (Fig. 9) and is synonymous with the so-called Sparganium emersum community (Herr 1984; Wiegleb and Herr 1984a; Herr et al. 1987).

Mapping of communities had to be restricted to the comprehensive types as the dominant growth form could be used for defining ecological differences between the geographical areas. Floristic composition is a much less stable element of vegetation than dominant growth form. Thus, the minor types, as defined above, are not mapped.

Classification into minor types was based on the use of partial data sets from more or less homogeneous geographical areas (Wiegleb 1983a). The data were also floristically relatively homogeneous. Species replacements were not considered changes in the actual character of vegetation when growth forms were similar. Thus, within each region, structurally similar vegetation types were produced that differ more or less in species composition but may be considered vicariant vegetation types.

Following this procedure, thirty-four minor types were identified from 330 relevees (Wiegleb 1983b). When the analysis includes all the relevees, it can be estimated that about 100 minor types can be characterized floristically and ecologically. Their protection value is defined as they are rare regarding a larger area but typical for a smaller region or a certain river system. This corresponds to the concept of the river (system) as a unique landscape unit.

ASSESSMENT

One very important character of wetland vegetation is its value for nature conservancy. The basic ideas for its assessment have been outlined elsewhere (Herr et al. 1978/79; Wiegleb 1984). An assessment is only possible if the conservation goals are clearly defined. Our approach to this objective was not to be numerically exact, and thus differs from more sophisticated approaches (e.g. Bauer 1971; Seibert 1980; Schuster 1980). Our objective was to protect all species. For this purpose habitat conservation is of special importance. Within this framework we used three different approaches which are outlined in Herr et al. (1989). The most important is assessment of vegetation by means of the regionally differentiated species deficit model. This model can only be applied in vegetation dominated by macrophytes and is based on the comprehensive information about spatial and temporal distribution of vegetation in Lower Saxonian water courses. The method is based on optimal species numbers and therefore considers potential natural vegetation. This approach is useful as similar habitat conditions will result in the development of comparable types of biocoenoses, as long as catastrophic and anthropogenic impacts are not too intensive. Additionally, not all species are considered to be of equal value, but are differentiated based on four criteria:

- occurrence of species typical for geographical area of river zone
- occurrence of species with supposed positive indicator value
- total abundance and vitality of the stands
- growth form diversity.

RESULTS AND CONCLUSIONS FOR LOWER SAXONY

The results of our assessment are represented in Fig. 10 and compiled in Table 1. We recognized three conservation classes:

I. Rivers with high value for protection; II. rivers with minor protective value, and III. rivers in which macrophytes are sparse or absent.

Table 1. Estimated distribution of ecological value classes (in percent of total river length) in relation to river systems (top part of table) and geographical areas (bottom part of table).

Conservation class	1	2	3
Rivers			
Elbe area	21	61	18
Weser area	5	48	47
Ems area	7	53	40
Vechte area	0	73	27
Coastal areas	8	50	42
Highlands and Lowlands			
Coastal marshes	6	30	64
Oldenburg-East Frisian plain	5	59	36
Stade diluvial plain	4	68	28
Ems-Hunte diluvial plain	10	59	31
Lüneburger Heide	22	70	8
Weser-Aller lowland	4	51	45
Boerde (loess) area	2	12	86
Osnabrück hill country	0	40	60
Weser-Leine highlands	1	34	65
Lower Saxony	1	52	40

In Table 1 we have summarized the data by geographical areas (Fig. 1) and drainage basins (Fig. 2). The Harz Mountains (Fig. 2) are not covered in this analysis because most water courses in that area belong to the cryptogamic region and bear no macrophytic vegetation. In total, only 8% of the surveyed rivers contain a vegetation of high conservation value, while 32% are of minor value with possibilities for restoration. Forty percent are of no value for nature conservation in the present state and restoration seems almost impossible at these sections. Regarding geographical areas and river systems a differentiation can be obtained. In the lowlands (Fig. 1), areas 2, 3, 4, 5 and 6 are formed by sandy deposits of the Saale-ice age. Most moraines are already leveled in this region and especially in the western part there are numerous bog and fen regions. The most valuable wetlands occur in the more or less natural water courses of the Lüneburger Heide (area 5) with its relatively sparse human settlement. Because the main rivers of this area belong to the Elbe system (Fig. 2) there are most class I scores reached. In the coastal marsh areas numerous species of high protective value occur, but only with sparse cover and reduced vitality. Therefore there are many class III sections in this area. In the other three diluvial plains (areas 2, 3, 4) and the Weser-Aller-lowland (area 6) most rivers belong to class II and may be restored to class I. South of these regions along the margins of the highlands only one percent of the river courses still bears a species-rich macrophytic vegetation of class I. This is the Boerde (loess) area which has a high density of human settlement, intensive agricultural and related industrial activities. In this part of the Weser-area a restoration of river systems seems impossible under the given circumstances. The southern highlands (areas 8.1, 8.2) contain sandstone and limestone areas. Because of continuous anthropogenic disturbances including high waters produced by hydraulic engineering only zero to one percent of the river courses still bear a species-rich vegetation of class one. Evaluation of water courses based on macrophyte vegetation is only one part of an overall assessment. Fauna, the geomorphological condition, and chemical and hydrological conditions must also be taken into account. The rivers also need to be examined as a landscape until which includes all of the tributaries and the watershed. From this perspective, Fig. 10 shows that most of the rivers and streams have been degraded over much of Lower Saxony. What type of actions must be taken in an area that is already highly degraded?

Initially, society must make the decision that in every geographical area and drainage area, natural water courses of different kinds must survive. Where they have already been severely damaged, measures must be taken to restore rivers to a condition which might have been found before intensive human influence took place (Dahl and Wiegleb 1984). These aims can only be achieved by an integrated system of river conservation areas. As a long-term perspective at least 10 to 12 percent of the total river length of the state are required for conservation. This minimum is assumed in analogy to the requirements of Heydemann (1980) for terrestrial systems. Taking the great diversity of river types in the state into account, about 20 percent are needed for effective conservation.

IMPLEMENTATION OF REGULATIONS FOR PURPOSES OF CONSERVATION

Results of ecological research currently have little influence on regional planning and development. This is especially true for conservation of water courses. We believe that some conservation can be attained by using existing laws.

In Lower Saxony the nature conservancy administration can utilize several laws. According to 12 (1) BNatSchG (Federal Nature Conservancy Act), a very time-consuming procedure which requires detailed preparation, a high degree of success is possible. According to 41 (2) NNatG (Lower Saxonian Nature Conservancy Act) "special conservation orders" can be provided for every place, where specifically protected plants and animals (according to Federal Species Conservation Act of 1980) are found. In addition every sort of "reed swamp" is protected according to 36 (3) NNatG. Different sorts of reeds frequently occur along Lower Saxonian rivers. Thus, hydraulic engineering devoted drainage problems would only be impossible, if those paragraphs were strictly applied. Unfortunately, there are also regulations that supercede those three noted above. We believe that the latter regulations have to be checked by the water authorities and that it should not be easy to ignore the conservation laws. Water authorities can be caused to carry out a more natural and biologically oriented management scheme.

Indeed, comprehensive management plans are required by 36b Federal Water Act (Wasserhaushaltsgesetz). But so far only water quality aspects have been subjects of planning. According to the law, habitats in the river and its floodplain, river utilization and land use in the adjacent areas should all be included in the planning. Thus far, however, the Federal Water Act has not been fully implemented.

Finally, the interference regulations (Eingriffsregelungen) according to 8 (2) BNatSchG can be used. This regulation requires that human interference must not adversely affect ecological conditions that can be described by

- the biological type of the water
- the function of the water with respect to physical, chemical, and hydrological properties, and
- the function of the water as an entity of distribution

If those effects are unavoidable, the interference has to be omitted. Just as mentioned above, the interference may take place in cases of higher necessity. But this necessity has to be gauged, too, and cost-benefit-calculations have to be required. A legal basis is not only given for the protection of the actual status but also for rehabilitation measures. In 24 BNatSchG, places where plants and animals may live in the future are also included in the protection concept. Thus, there exists a number of regulations which enable people engaged in nature conservation to assert their interests against other groups mainly interested in the exploitation of nature.

ACKNOWLEDGEMENT

The studies were financially supported by the Niedersächsisches Landesverwaltungsamt - Fachbehörde für Naturschutz.

REFERENCES

Bauer, H.J. 1971. Landschaftsökologische Bewertung von Fliessgewässern. Natur und Landschaft 46: 277-282.

Dahl, H.J. and Wiegleb, G. 1984. Zielvorstellungen eines zukünftigen Fliessgewässerschutzes. Jahrbuch Naturschutz Landschaftspflege 34: 26-65.

Grube, H.J. 1975. Die Makrophytenvegetation der Fliessgewässer in Süd-Niedersachsen und ihre Beziehungen zur Gewässerverschmutzung. Archiv für Hydrobiologie Supplement 45: 386-456.

Herr, W. 1984. Die Fliessgewässervegetation im Einzugsgebiet von Treene und Sorge. Mitteilungen der Arbeitsgemeinschaft Geobotanik Geobotanik Schleswig-Holstein und Hamburg 33: 77-117.

Herr, W., Todeskino, D. and Wiegleb, G. 1978/79. Untersuchungen über die Schutzwürdigkeit und Regenerierbarkeit des Neudorfer und Stapeler Moores (Landkreis Leer, Reg.-Bez. Weser-Ems). Oldenburger Jahrbuch 78/79: 453-492.

Herr, W. and Wiegleb, G. 1985. Die Potamogetonaceae niedersächsischer Fliessgewässer. Teil II: Schmalblättrige Arten. Goettinger Floristische Rundbriefe 19: 2-16.

Herr, W., Todeskino, D., and Wiegleb, G. 1989. Übersicht über Flora und Vegetation der niedersächsischen Fliessgewässer unter besonderer Berücksichtigung von Naturschutz und Landschaftspflege. Naturschutz u. Landschaftspflege in Niedersachsen: 18:145-283.

Heydemann, B. 1980. Die Bedeutung von Tier- und Pflanzenarten in Ökosystemen, ihre Gefärhdung und ihr Schutz. Jahrbuch Naturschutz Landschaftspflege 30: 15-87.

Londo, G. 1984. The decimal scale for releves of permanent quadrats. pp. 45-50. In R. Knapp, Sampling methods and taxon analysis in vegetation science. Junk, The Hague, The Netherlands.

Roll, H. 1938. Allgemein wichtige Ergebnisse für die Pflanzensoziologie bei der Untersuchung von Fliesswasser in Holstein. Feddes Repertorium Beihefte 101: 108-112.

Schuster, H.J. 1980. Analyse und Bewertung

von Pflanzengesellschaften im Nördlichen Frankenjura. Dissertationes Botanicae 53. Vaduz.

Seibert, P. 1980. Ökologische Bewertung von homogenen Landschaftsteilen, Ökosystemen und Pflanzengesellschaften. Ber. ANL 4: 10-23.

Weber, H.E. 1976. Die Vegetation der Hase von der Quelle bis Quakenbrück Naturwissenschaftlichen Mitteilungen 4: 131-190.

Weber-Oldecop, B.W. 1970. Wasserpflanzengesellschaften im Östlichen Niedersachsen I. Internationale Revue der Gesamten Hydrobiologie 55: 913-967.

Weber-Oldecop, B.W. 1971. Wasserpflanzengesellschaften im oestlichen Niedersachsen II. Internationale Revue der Gesamten Hydrobiologie 56: 79-122.

Wiegleb, G. 1981. Probleme der syntaxonomischen Gliederung der Potametea. pp. 207-249. In H. Dierschke, ed. Syntaxonomie. Cramer Vaduz.

Wiegleb, G. 1983a. Recherches méthodologiques sur les groupements végétaux des eaux courantes. Colloques Phytosociologiques 10 (végétations aquatiques): 69-83.

Wiegleb, G. 1983b. A phytosociological study of the macrophytic vegetation in running waters in Western Lower Saxony (Federal Republic of Germany). Aquatic Botany 17: 251-274.

Wiegleb, G. 1984. Makrophytenkartierung in niedersächsischen Fliessgewassern - Methoden, Ziele und Ergebnisse. Informationen zu Naturschutz und Landschaftspflege 4: 109-136.

Wiegleb, G. and Herr, W. 1983. Taxonomie und Verbreitung von Ranunculus subgenus Batrachium in niedersächsischen Fliessge- wässern unter besonderer Berücksichtigung des Ranunculus penicillatus-Komplexes. Göttinger Floristische Rundbriefe 17: 101-150.

Wiegleb, G. and Herr, W. 1984a. Zur Entwicklung vegetationskundlicher Begriffsbildung am Beispiel der Fliessgewasservegetation Mitteleuropas. Tüxenia 4: 303-325.

Wiegleb, G. and Herr, W. 1984b. Die Potamogetonaceae niedersächsischer Fliessgewässer, Teil I: Breitblättrige Arten. Göttinger Floristische Rundbriefe 18: 65-86.

VEGETATION AND LAND USE IN THE LUZNICE RIVER FLOODPLAIN AND VALLEY IN AUSTRIA AND CZECHOSLOVAKIA

K. Prach, S. Kucera and J. Klimesova
Institute of Botany, Czechoslovak Academy of Sciences,
Dukelská 145, CS-379 82 Trebon, Czechoslovakia

ABSTRACT

Vegetation in the floodplain and valley of a middle-size river in Central Europe was analysed and related to past and present types of land use. A general phytosociological description and more detailed analyses of thirty-three cross-section transects were used to interpret vegetational patterns in relation to intensity of anthropogenic changes in the landscape.

INTRODUCTION

River floodplains are highly dynamic ecosystems that are influenced by recent and past alterations in land use of wetlands, terraces that border them, and in the adjacent landscape including the whole catchment area. Ecosystems of floodplains thus may be considered as a focal points for impacts imposed by the surrounding landscape. We agree that vegetation is a good indicator of these interactions.

The middle-sized Luznice River (Lainsitz in German) is a natural axis of the Trebon basin in South Bohemia. It traverses a small part of the Weitra Region in Lower Austria. In the Trebon Biosphere Reserve, declared in the Czechoslovakian part of the Trebon basin in 1977, long term investigations within IBP and MAB Programmes have been carried out, dealing with the ecology of wetlands attached to water bodies (Dykjová and Kvet 1978; Jeník and Kvet 1984). This is one of the first papers on the ecological project "Floodplain of the Luznice River", sub-project within the MAB project No. 87, "The Trebon Project - Role of Wetlands in the Temperate Forest Biome". The paper gives an overview of the vegetation of the bottom and flanks of the valley along the entire length of the river. The field work was mostly conducted in 1987. Three basic questions were addressed:

(a) Does vegetation change directionally along the geographic length of the river?

(b) How does the vegetation reflect the pattern of land use in the floodplain and in its surroundings?

(c) How well does the part of the river valley selected for detailed ecological study represent the regional vegetation and land use?

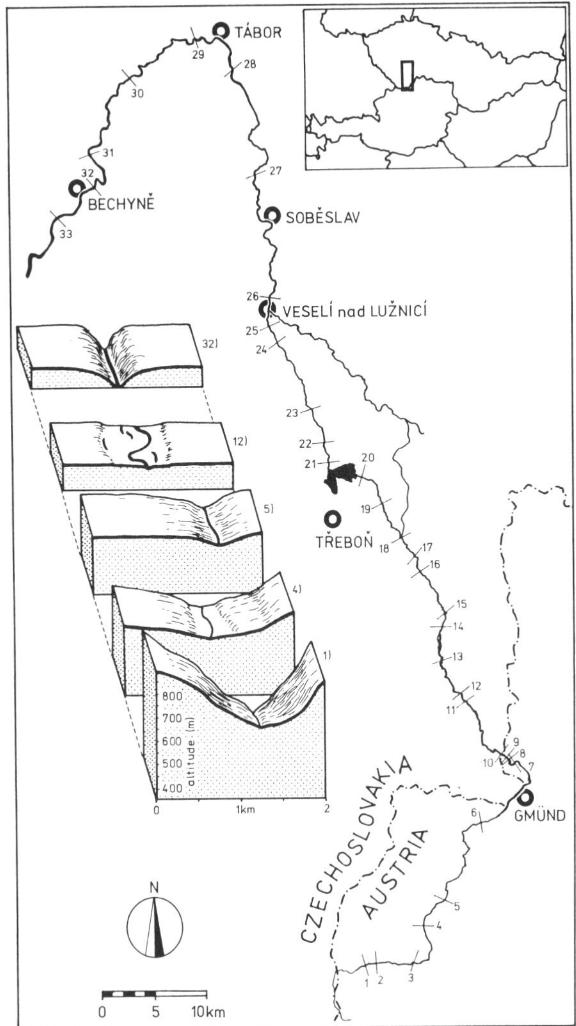

Figure 1. Map of the Luznice River showing locations of transects and cross-section diagrams of typical parts of the valley.

D. F. Whigham et al. (eds.), Wetland Ecology and Management: Case Studies, 117–125.

Figure 2. The Luznice River Valley in the Novohradské Hory Hills. (Friewald). (River Section 1).

STUDY AREA

The Luznice River, total length 200 km, originates in the forested Central European hills of the Novohradské hory (Friewald in Austria) at the altitude of 990 m. It empties into the Vltava River at an altitude of 347 m (Figs. 1 and 2) and drains 4225 km^2 of Central European Territory. Basic geographical information on four distinct sections of the river is given in Table 1, partly using criteria after Haslam (1987). The general vegetation characteristics of each river section follows, using the vegetation units of the Zürich-Montpellier school of phytosociology. The units are mostly given after Oberdorfer (1957), considering adaptations of Barkman et al. (1986). The Syntaxa abbreviations: cl. - classe, ord. - ordo, all.- alliance, subal.- suballiance, ass. - association.

High Upland Spring Section

The Novohradské hory hills, where the Luznice River originates, are floristically related to the Alps. In the predominantly forested region, springs and stream valleys, including those of the Luznice River, are suitable habitats for the persistance of subalpine species, (e.g., Veratrum album and Ranunculus aconitifolius). Human colonization did not occur until the period 1600-1900 A.D., so that the human impact on the flora and vegetation has been comparatively small; mostly only quantitive changes have occurred in the flora.

The headwaters of the river are located in Norway-spruce plantations which occupy areas that were originally dominated by beech-forests (The Fagion all.). Spring communities of the Cardamino-Montion all. are frequent there. The adjoining plateau is covered by fragments of original Norway-spruce communities of the Vaccinio-Piceion all., in openings with broad-leaved herb communities of the Calthenion subal. Downstream the Luznice River turns to a rapid mountain creek which is fringed by a mosaic of tall herb communities of the Adenostylion All., of various spring communities of the Montio-Cardaminetea cl., and of alder carrs of the Alnenion glutinoso - incanae subal. (ass. Arunco sylvestris-Alnetum glutinosae and ass. Piceo-Alnetum).

Austrian Upland Section

This river section has been strongly modified by human impact since about 1000 A.D. Gradual clearing of the forest was accompanied by an invasion of ancient weeds (archeophytes) and by an expansion of species and communities typical of both wet and dry meadows (Ricek 1982). The last century is remarkable for only moderate disturbance and extirpation of the local flora. The river valley offers suitable habitats to all mentioned groups of species, thus being the most diverse landscape element here.

Wet meadows of the Calthenion subal. prevail in the floodplain and pass into mesophilous meadows of the Arrhenatherion all. Species rich meadows of the latter syntaxon with dominant Holcus lanatus are a distinct vegetational type here. Fragments of short grass communities of the Violion caninae Schwikerath 1944 all occur on dry flanks of the Valley (Fig. 3). All meadows are regularly mown or grazed; and woody species are restricted to a narrow riparian belt along the stream (the alder carr of the Alnenion glutinoso-incanae subal.) or to steep slopes (originally beech forests of the Luzulo-Fagenion subal. and, exceptionally, oak woods of the Genisto germanicae-Quercion Neuhäusl and Neuhäuslova-Novotná 1967 all..

The Trebon Basin Section

The colonization of the Trebon basin started comparatively late, 12th century, and the further development of this flat, poorly drained and peaty landscape differed from that in neighboring regions in many characteristic features (Jenik and Kvet 1984; Jankovski 1980). In the broad floodplain of the Luznice River, a diverse set of wetlands has developed along streams, canals and fishponds, depending on the water table dynamics. (Figs. 4 and 5). Some species typical of riverine marshes implicate phytogeograpical relations to the Danube plain (Rorippa amphibia, Gratiola officinalis, Fritillaria meleagris). Some relic species of the former open periglacial landscape are noteworthy (Pulsatilla vernalis, Anthericum

Table 1 - Physiographic Characteristics of the Luznice River Sections

Characteristics	River Sections 1	2	3	4
River kilometers	200-185	185-160	160-56	56-0
River elevation (meters above sea level)	990-670	670-500	500-403	403-347
Topograpical class of landscape	High Upland	Upland	Lowland	Upland
Mean annual temperature (^{o}C) and precipitation (mm)	4.9; 901[1)]	6.6; 695[2)]	7.8; 627[3)]	7.3; 602[4)]
Prevailing rock type	Granodiorite	Granodiorite Paragneiss	Sand, clay	Metamorphites
Fall from hill tops to stream channel (meters)	Up to 250	100	10	120
Average slope of channel (%)	2.1	1.8	0.93	1.0
River-bed width (meters)	1-3	3-8	8-30	30-40
Average long term discharge + ($m^{3}\cdot s^{-1}$)	less than0.5	0.5-3.0	3-20	20-25
Trophic status	Oligotrophic	Oligotrophic Mesotrophic	Eutrophic	Eutrophic
Width of floodplain (m)	±0	50-150	150-1000	0-50
Transects numbers as shown on Fig. 1.	1	2-5	6-27	28-33

1)Karlstift, 950 meters above sea level; 2) Weitra, 580 meters; 3) Trebon, 433 meters; 4)Tábor, 441 meters (all data are for the period 1901 - 1950).

+ Based on data of the Hydrometeorlogical Institute at Ceské Budejovice

ramosum, Daphne cneorum) usually growing on dry convex sandy slopes, or Spiraea salicifolia typical of wet sites in the floodplain.

In the less cultivated and more eutrophic parts of the floodplain, various tall-graminoid marshes of the Phragmiti-Magnocaricetea cl. and fragments of short graminoid communities of the Scheuchzerio-Caricetea fuscae cl. have developed together with productive meadows of the Alopecurion all. and of the Calthion all. The river is fringed by the Salicion triandrae Muller, Gors 1958 all. some invasive willow stages of the Salicetalia auritae Doing 1962 ord. can also be noted. Certain segments of the floodplain are covered by swamp-alder communities of the Alnion glutinosae all. Species poor communities of the Callitricho-Batrachietalia Passarge 1978 ord. occupy the river bed and those of the Potametea cl. occur in permanent water pools in oxbows and backwaters. Short-grass communities of the Violion caninae all., and the Festuco-Sedetalia ord. are typical of dry slopes and tops of terraces. Large areas on terraces are covered by forest plantations, usually with dominant Pinus sylvestris (originally mixed and deciduous woods of the Dicrano-Pinion Libbert 1933 all. and the Genisto germanicae-Quercion all.).

Figure 3. The narrow floodplain with regularly mown meadows in the Austrian upland. (River section 2).

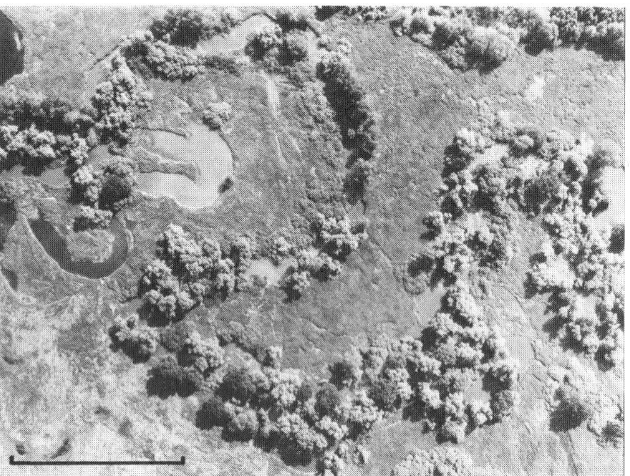

Figure 4. An aerial view on a part of the Luznice River floodplain in the Trebon Basin (River section 3) with the meandering river, oxbows and backwaters, and mostly unmanaged marshes. Bar in lower corner corresponds to 100 m distance.

Figure 5. An early spring flood in the river section 3.

The Canyon-Like Downstream Section

The last section (Fig. 6), in the narrower and deeper valley, is characterized by the occurrence of ecologically constrasting plant communities and species. Various originally mountain species (Doronicum austriacum), Cardaminopsis hallerii, Soldanella montana) are often scattered there in fragments of a Norway-spruce forest concentrated at the bases of shaded flanks. Heliophilous species can be noted as relics of vegetation of periglacial or earlier periods of the Holocene (Alyssum saxatile, Arctostaphylos uva-ursi, Festuca pallens, Hieracium cymosum, Pinus sylvestris, Quercus petraea, Betula pubescens). Elements of mesophilous deciduous forests also characterize the last river section (e.g. Galeobdolon luteum, Galium sylvaticum, Hepatica nobilis).

The river is skirted by alder communities of the Alnenion glutinoso-incanae suball. Originally species-rich meadows of the Calthenion suball. and of the Arrhenatherion all. are undergoing the processes of ruderalization associated with recreational activities on the river valley, and as a consequence of absence of mowing. Cultivated forests of Norway spruce and Scotch pine prevail on the flanks, together with small fragments of original woody communities of the Tilio-Acerion Klika 1955 all., Carpinion all., Genisto germanicae-Quercion all., Quercion pubescenti -petrae all. and Dicrano-Pinion all., and small patches of open relic vegetation on shallow soils (the Alysso-Festucenion pallentis suball., the Prunetalia ord.).

Human colonization of the surrounding landscape started as early as the Neolithic period (ca. 4000 B.C.), but the river valley itself was largely saved from ancient human impact. At present, ruderalization is most evident at the bottom and at the contacts of the flanks and agrocoenoses prevailing on the adjoining peneplane. An ecological contrast between the river valley and the surrounding landscape is conspicuous.

METHODS

For a quantitative evalutation of the vegetation along the cross-sectional transects, it was necessary to delimit simpler and less rigid vegetation units than are those of the Zürich-Montpellier (Z-M) system. Various transient, successional, degradational and ruderalized stages as well as man-made ecosystems, which are only rarely considered by the Z-M system, had to be classified. Thus, we had to apply rather different criteria to a unit definition, predominantly using dominant species, and in some cases also considering background ecological factors, including human activities (see Table 2).

Figure 6. The Luznice River in the canyon-like downstream section. (River section 4).

The transects started and ended at the top edges of the flanks or at the top of the first terrace (Fig. 1). In several cases, when it was difficult to define the tops of flanks or terraces, we used an arbitrary end of a transect: 500 m from the river bank.

Segments of the transects occupied by each vegetation unit were measured (m) and briefly characterized as to moisture conditions, soil and surface relief. Vegetational diversity was calculated using the Shannon Index (H_V):

$$H_V = -\sum_{i=1}^{n} \frac{l_i}{l} \log_2 \frac{l_i}{l}$$

Where l_i means the length of a transect occupied by a vegetational unit i and l is total length of a transect; n is the number of vegetational units distinguished along a transect.

Data were arranged into a table with transects as columns and vegetational units as rows. The untransformed data were processed by the detrended correspondence analysis (Hill 1979).

RESULTS

The geomorphology and land use of the river valley can be characterized by the occurrence of forests, regularly mown meadows, and various artificially disturbed sites expressed as per cent of the length of each transect (Fig. 7). Participation of all vegetation - ecological units distinguished along the transects in the particular river sections is given in Table 2, together with other information on the length of transects, number of units and vegetation diversity. Average participation of the vegetation-ecological units in the whole set of transects analysed is given in the last column of the table.

Arable land is the most common unit, followed by spruce forests, and grassland communities dominated by Alopecurus pratensis or Baldingera arundinacea. Riparian Urtica dioica communities, willow carr, and various ruderal communities are also widespread along the entire river length, except for the first few kilometers. Most of these communities are indicative of changes due to human impact.

Results of the DCA ordination (Fig. 8) corroborate our field observations of the river valley vegetation. Transects were arranged along the first ordination axis which reflects the intensity of land use (degree of land transformation) increasing from forestry on the right to sand-gravel excavation. The three clusters represent the three main geomorphological types of the river valley: (A) Canyon-like (disregarding location of the transects along the river); (B) Areas where the floodplain is narrow; and (C) In the Trebon basin where the floodplain is broad. The relationship between geomorphological settings and land use are evident. Steep areas are predominantly forested, the homogeneous but narrow floodplain in the Austrian upland is convenient for intensively mown meadows while the large and diversified flat floodplain near Trebon is subjected to various human activities, even large scale excavations of sand and gravel. Large tracts of former meadows in the Czechoslovak part of the river valley have been abandoned and are undergoing the processes of ruderalization and expansion of woody species. This is the reason for large dissimilarities between the "meadow" transects in the cluster B and those in the cluster C in Fig. 8.

The results of the direct gradient analysis (Fig. 7) and those of the indirect gradient analysis (Fig. 8) are mutually comparable and complementary.

DISCUSSION

A detailed quantitative description of the vegetation along a whole river course is difficult

Table 2 - Particpation of vegetation-ecological units along the transects in particular river section (in percent of total length of the transects). com. = community A = average.

No. of sections	1	2	3	4	A
No. of transects	1	2-5	6-27	28-33	
Total length of transects /m/	1000	617	11452	2302	
1. Running Waters	0.1	3.0	2.9	8.7	3.6
2. Oligotrophic permanent pools	.	.	3.5	.	2.6
3. Other permanent pools	.	.	1.2	0.04	0.9
4. New drifts and emergent streambars	.	.	0.2	.	0.1
5. Emergent muddy bottoms of pools with ephemeral communities	.	.	0.1	.	0.06
6. Rorippa amphibia com.	.	.	0.5	.	0.4
7. Glyceria fluitans com.	.	3.7	0.03	.	0.2
8. Carex vesicaria com.	.	0.5	0.7	.	0.5
9. Carex rostrata com.	.	1.0	.	.	0.04
10 Carex gracilis com.	.	.	3.4	.	2.5
11. Carex buekii com.	.	.	0.1	.	0.3
12. Carex juncella com.	.	.	0.1	0.8	0.1
13. Phragmites australis com.	.	.	0.1	1.3	0.2
14. Glyceria maxima com.	.	.	2.0	0.1	1.5
15. Filipendula ulmaria com.	.	0.5	0.2	2.3	0.5
16. Scirpus sylvaticus com.	.	2.0	0.4	0.9	0.5
17. Juncus filiformis com.	.	.	0.1	.	0.07
18. Juncus effusus com.	.	.	0.1	.	0.1
19. Calamagrostis canescens com.	.	.	0.6	.	0.5
20. Calamagrostis phragmitoides com.	.	1.3	.	.	0.06
21. Baldingera arundinacea com.	.	0.6	9.2	0.8	7.0
22. Riverine Urtica dioica com.	.	.	2.9	0.4	2.3
23. Deschampsia cespitosa meadows	.	.	1.4	0.2	1.1
24. Alopecurus pratensis meadows	.	0.5	8.8	3.1	7.1
25. Molinia coerulea meadows	.	.	0.5	.	0.4
26. Species rich mesophilous meadows with Holcus lanatus	.	44.0	0.2	6.0	2.8
27. Other mesophillous meadows	.	16.0	1.0	0.3	1.4
28. Repeatedly reclaimed meadows	.	.	0.6	.	0.5
29. Short grass meadows on drier sites	.	.	0.6	.	0.5
30. Carex brizoides grasslands	.	0.5	0.6	.	0.5
31. Willow carrs	.	5.0	3.6	1.4	3.1
32. Spiraea salicifolia shrubs	.	.	0.6	.	0.4
33. Alder carrs	.	.	3.6	2.7	3.0
34. Mountain spruce-alder com.	1.5	.	.	.	0.1
35. Quercus robur woods on wetter sites	.	.	6.5	1.3	5.0
36. Quercus robur and Q. petraea woods on dryer sites	.	1.1	.	14.2	2.2
37. Rich mesophilous mixed woods	.	18.0	0.1	6.5	1.1
38. Acidophilous beech forests	42.1	.	.	.	2.8
39. Scotch pine forests	.	.	2.9	15.6	4.5
40. Semi-natural Norway spruce forests	3.3	.	.	.	0.2
41. Norway spruce plantations	49.8	.	3.3	26.2	9.7
42. Shaded spring communities	0.5	.	.	.	0.03
43. Open rock communities	.	.	.	1.2	0.2
44. Wet forest fringes	.	.	.	0.3	0.04
45. Forest fringes on drier sites	1.3	0.5	0.1	0.1	0.1
46. Forest mantles on drier sites	1.3	.	0.1	1.7	0.4
47. Newly disturbed ruderal sites	.	.	2.5	0.3	1.9
48. Ruderal com. of tall perennial plants	0.5	0.2	4.2	0.9	3.3
49. Trampled sites	0.2	.	0.2	0.6	0.3
50. Arable land	.	.	28.9	0.9	21.7
51. Buildings, roads, etc.	0.5	0.6	0.1	0.3	0.1
Mean number of units per transect	11.0	8.0	12.1	12.3	10.9
Mean veget. diversity per transect	1.6	1.7	2.4	2.5	2.1

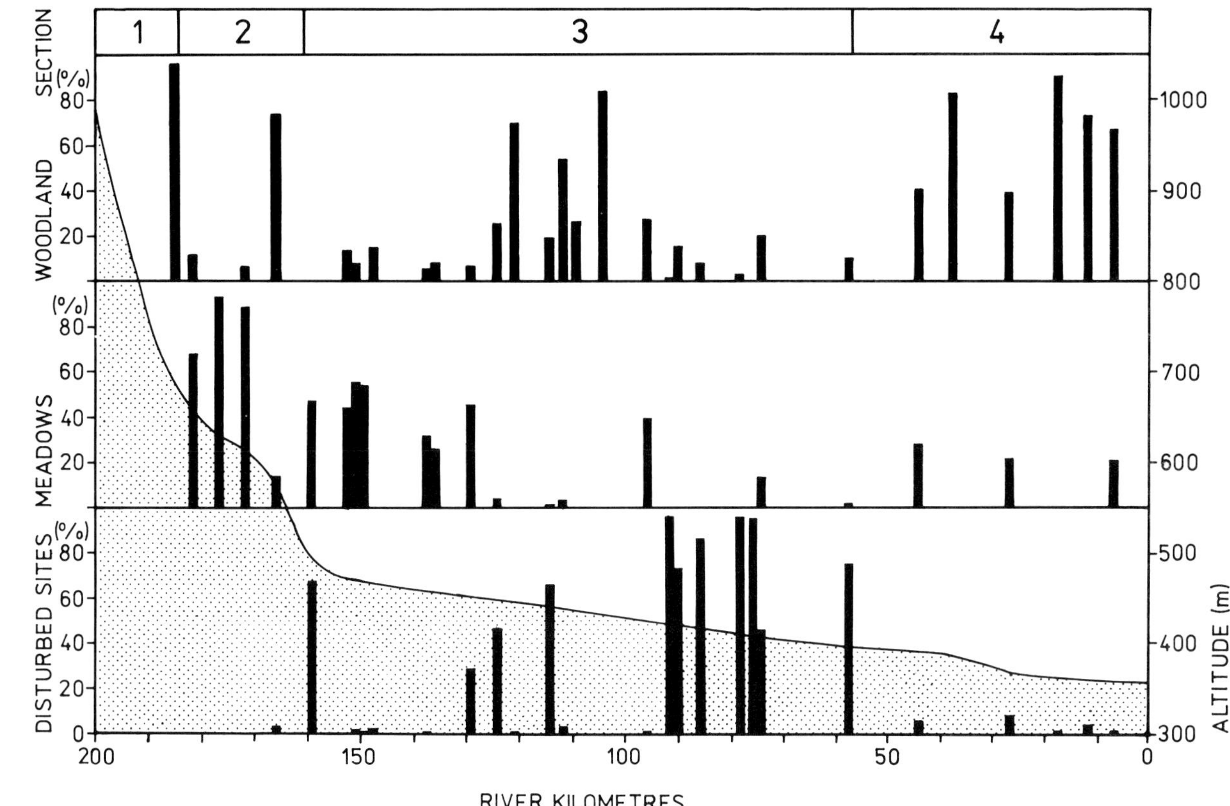

Figure 7. Percent cover (in per cent of total length of each transect) of woodlands, regularly mown meadows, and various artificially disturbed sites (arable land, roads, buildings, quarries, etc.) down the river. Lengthwise, the river profile is depicted and basic river sections are marked (see Table 1).

because of the complexity of ecological gradients and land use (Haslam 1987; Wiegleb 1988). Such descriptions have usually been made only in a stream, in a narrow vegetation belt along a stream, within one type of vegetation, or within the distributional range of particular species (Kopecky 1970; Weigleb 1988). The occurrence of particular species was evaluated along the cross-section transects by Curry and Slater (1986). However, detailed vegetation analyses across a river valley were usually made only at one or a few spots along a river (e.g., Menges and Waller 1983; Metzler and Damman 1985). Despite methodological and conceptual limitations, description of the vegetation in and along a river can yield valuable information on a whole region, its natural conditions, history and present land use, implicating social and even political aspects (Whitton 1984; Décamps 1984; Mitsch and Gosselink 1986). Two approaches were used in this paper to evaluate the riverine vegetation: the "classical" Central European phytosociological approach and gradient analysis. The former was more efficient in the general description of vegetation and the latter in detailed vegetation analyses along the cross section transects. In the latter approach, we used pragmatically defined units which, however, can be optimum only in the vegetation analysis within the given river system. A repeated analysis along the same transects will be possible in the future because they were permanently marked.

Land use markedly differs in the Austrian and Czechoslovakian part of the river valley and its surroundings. In Austria, most floodplains are regularly mown and without enormous fertilizing. The adjoining landscape is not evidently polluted. In Czechoslovakia, large areas of the floodplain are not cultivated while the stream is polluted by waste waters and by run-off from heavily fertilized fields and meadows both in the floodplain and in surrounding areas (Drbal et al. 1988).

Floodplain meadows seem to be the best indicators of land use. They respond rapidly to changes associated with direct management as

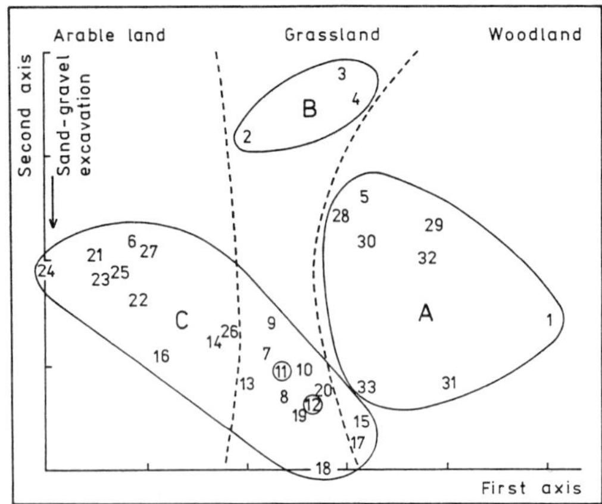

Figure 8. Ordination of vegetational transects based on occurrence of vegetation-ecological units in meters. Clusters represent transects in three distinct geomorphological areas within the valley: (A) canyon-like; (B) areas with a narrow floodplain in the upland; (C) areas with a wide floodplain in the Trebon Basin. Prevailing land use is depicted along the top of the diagram. Transect no. 19 is dominated by woodland, it differs greatly in the third axis score; transects no. 11 and 12 (encircled) are currently being studied in greater detail. See Fig. 1 for location of transects.

well as without management in surrounding areas that are hydrologically linked to the floodplain. In the Luznice River floodplain, meadow management is often characterized by decreasing intensity of mowing and by direct enormous application of manure. In those situations, one species usually becomes dominant and diversity decreases. These phenomena correspond with theoretical expectations when diversity is decreasing at high levels of nutrient input and low disturbance intensity, i.e., the absence of mowing in our case (see Grime 1979; Tilman 1988). The nutrients: disturbance loss gradient of Tilman (1988) is a useful tool for interpretations of present vegetational phenomena in the floodplains and experiments to test these ideas are in progress (Srutek et al. 1989). Baldingera arundinacea, Urtica dioica, and Glyceria maxima become increasingly important when the input of nutrients is increased. On the other hand, all these species are susceptible to the intensity of the mowing regime, especially Urtica dioica. The decline of less robust and less competitive species is enhanced by the accumulation of litter in the absence of mowing.

Species rich meadows with Holcus lanatus, potentially the most common vegetation type in the floodplain (no.26 in Table 2) were noticed on large areas only in the Austrian part of the valley (See also Bálatová-Tulácková and Hübl 1985). In the Czechoslovakian part, this vegetation type is now present only in small fragments preserved from the time of the former floodplain and landscape management.

In cultivated parts of the floodplain, woody species are restricted only to a narrow strip along the stream or, exceptionally, around oxbows and backwaters. They can expand on unmown sites, but their establishment is often retarded by a dense cover of the highly competitive and productive herbaceous species (Dykyjová and Kvet 1978). Thus, some wetlands now present on uncultivated parts of the floodplain may be considered to be disclimax (Whittaker 1974), consequently they are not attractive sites for agriculture.

It is generally evident, that optimal management of the floodplain should be that to which the whole system had become adapted in the course of several centuries. Regular mowing and application of low doses of fertilizers appears to be appropriate on sites that are not regularly flooded. On regularly flooded sites, the deposited silt replaces the nutrients removed in the hay.

CONCLUSIONS

The results presented here gave answers to questions posed in the introduction:

(a) The vegetation changes are not simply unidirectional in a downstream direction along the Luznice River and they exhibit no clear gradient along the whole river course. Altitude, local geomorphology of the valley, and both previous and present land use interfere and determine the character of each particular habitat. Phytogeographical relations are partly responsible for the development of a particular vegetation type.

(b) Land use, which largely controls the present vegetation in the floodplain, is predominantly determined by geomorphological conditions in different parts of the valley and of the adjoining landscape. Specific historical, social, and even political factors, however, also play an important role. In Austria, nearly the whole flat floodplain is regularly and adequately mown, which is conducted in association with a low artificial nutrient input in species rich mesotrophic meadows. In the Czechoslovak part of the floodplain, we can see a successional gradient from regularly mown meadows to marshes unmanaged for up to 40 years. Many former meadows have changed to willow stands or to monotonous stands of competitive herbaceous species being promoted by a high level of nutrient input and by the absence of mowing.

(c) The part of the Luznice river valley, which has been selected for detailed ecological study (see Drbal et al. 1988), is representative of the large floodplain with predominant grassland occurring along the naturally meandering river, Agricultural activities will be decisive for the further development of this area (i.e, regularity and extent of mowing, intensity of artificial fertilization and of mechanical disturbance of the soil surface by heavy machines). If other more destructive measures are adopted (large scale sand and gravel excavation, channelization of the river), the impacts will be more severe there causing the river and it's floodplain to become further removed from natural geomorphic and hydrologic controls.

ACKNOWLEDGMENTS

We thank J. Jeník and J. Kvet for their helpful comments.

REFERENCES

Balátová-Tulácková E., and Hübl, E. 1985: Feuchtbiotope aus den nordoestlichen Alpen und aus der Böhmischen Masse. - Angewandte Pflanzensoziologie 29:1-131.

Barkman, J.J., Moravec, J., and Rauschert, S. 1986: Code of phytosociological nomenclature (2nd editon). Vegetatio 67:145-195.

Curry, P., Slater, F.M. 1986. A Classification of river corridor vegetation from four catchments in Wales. Journal of Biogeography 13:119-132.

Décamps, H. 1984. Towards a landscape ecology of river valleys pp. 163-178. In Cooley, J.H. and Golley, F. G., eds. Trends in ecological research for the 1980's. Plenum, New York, NY, USA.

Drbal, K., Jeník, J., and Prach, K., eds. 1988. Ecological Project "The floodplain of the Luznice River." (in Czech with English summary). Sborník, VSZ Ceské Budejovice 5(2): 1-158.

Dykyjova, D. and Kvet, J., eds. 1978. Pond littoralecosystems. Structure and functioning. Ecological Studies 28. Springer Verlag, Berlin, Germany 464 p.

Grime, J.P. 1979. Plant strategies and vegetation processes. J. Wiley, Chichester, U.K. 222 p.

Haslam, S.M. 1987. River Plants of Western Europe. Cambridge University Press, Cambridge, U.K. 512 p.

Hill, M.O. 1979. Decorana, a Fortran Programme for Detrended Correspondence Analysis and Reciprocal Averaging. Cornell University Press, Ithaca, NY, USA. 52 p.

Jankovska, V. 1980. Paläobotanische Rekonstrucktion der Vegetationsentwicklung im Becken Trebonska Panev Während des Spätglazials und Holozäns. Academia Press, Praha, Czechoslovakia. 151 p.

Jeník, J. and Kvet, J. 1984. Long-term research in the Trebon biosphere reserve, Czechoslovakia. pp. 437-459. In di Castri, F., Baker, F.W.G. and Hadley, M., eds. Ecology in practice, Part 1: Ecosystem management. Tycooly, UNESCO, Paris, France.

Kopecky, K. 1970. Neophytes in river-bank communities along the Orlice River in Northeast Bohemia (in Czech with English summary). Studia CSAV, Academia Praha, Czechoslovakia. 1970/77: 97-106.

Menges, E.S. and Waller, D.M. 1983. Plant strategies in relation to elevation and light in floodplain herbs. American Naturalist 122:454-473.

Metzler, K.J. Damman, A.W.H. 1985. Vegetation patterns in the Connecticut River flood plain in relation to frequency and duration of flooding. Nature Canada. 112: 535-547.

Mitsch, W.J.and Gosselink, J.G. 1986. Wetlands - Van Nostrand Reinhold Comp., New York, N.Y, U.S.A. 539 p.

Oberdorfer, E. 1957. Süddeutsche Pflanzengesellschaften. - Jena G. Fischer, 564 p.

Ricek, W. 1982. Die Flora der Umgebung von Gmünd im nieder österreichischen Waldviertel. Abhandlungen des Zoologische-botanische. Gesellschaft Österreich 21: 1-204.

Srutek, M., Bauer, V., Klimes L. and Pinosova, J. 1988. Ecology of economically important plant species in the Luznice River floodplain (in Czech with English summary). Sborník VSZ Ceské Budejovice 5(2): 105-118.

Tilman, D. 1988. Plant strategies and the dynamics and structure of plant communities.- Princeton University Press, Princeton, NJ, USA. 360 p.

Whittaker, R.H. 1974. Climax concepts and recognition pp.139-154. In Knapp, R., ed. Vegetation dynamics. W. Junk Publisher, The Hague, The Netherlands.

Whitton, B.A. 1984. Ecology of European Rivers. Blackwell Science Publisher, Oxford, U.K. 644 p.

Weigleb, G. 1988. Analysis of flora and vegetation in rivers: Concepts and Applications. pp. 311-340. In Symoens, J.J., ed. Vegetation of inland waters. Handbook of Vegetation and Science 15/1. Kluwer, Dordrecht, The Netherlands.

DANUBE BACKWATERS AND THEIR RESPONSE TO ANTHROPOGENIC ALTERATION

H. Löffler
Dept. of Limnology, University of Vienna
1090 Wien, Althanstr. 14, Austria

Abstract

Ecological conditions and species richness of biotic communities were studied in the floodplains of two relatively comparable sections of the river Danube in Austria: (a) an unaltered section near Hainburg (downstream from Wien), but where the construction of a power plant had been proposed, and (b) an altered section near Altenwörth (upstream from Wein), where a power plant has already been constructed. The observed differences in sedimentation rate, chemical composition of water, and species composition of flora and fauna are only partly due to differences in the quality of inflowing surface water and upwelling groundwater. Most of the differences reflect the different degrees of human impact upon the two sites. The unaltered site (a) is much richer in species and requires conservation as a unique example of a riverine landscape in Europe. The proposed power plant therefore must not be built there.

INTRODUCTION

The European riverine landscape has suffered severe loss by regulation of the majority of streams and their use for power stations. The Rhine, once flowed through about 1,000 km^2 of alluvial forests and associated floodplain features, such as evorsion, lateral, lateral levee lakes and oxbows. The Rhine is at present bordered by less than 75 km^2 of alluvial forests. Similarly, the riverine landscape of the Danube has been greatly reduced by the regulation of the river during the second half of the last century, by the construction of power plants, and by land reclamation (Löffler et al. 1976). The last portion of alluvial landscape - and at the same time the largest in Europe of some 80 km^2 - extends between Wien and the Austrian-Czechoslovakian frontier and is now threatened by the planned construction of a power station near Altenwörth (Fig. 1). This selection was made since both localities belong to the Pannonian Province and have similar climatic conditions (Fig. 1).

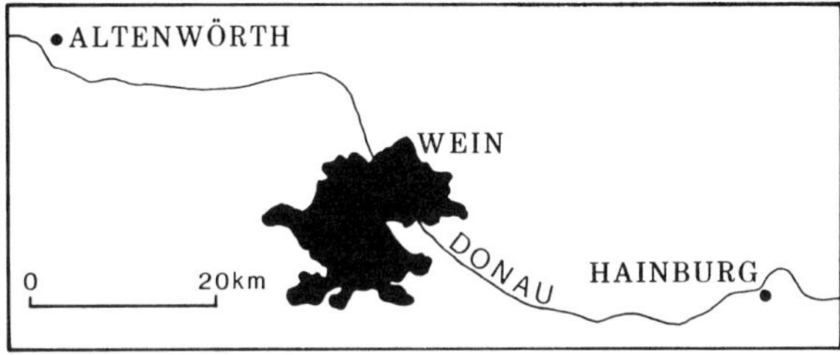

Figure 1. Location of Altenwörth Wein and Hainburg on the Donau River.

Apart from the power station, the complete separation of the Danube from the alluvial plain above Altenwörth, which prevents major changes of water level in the plain and rinsing of alluvial gravel from below, is the only important feature that differs from the undisturbed site east of Wien. Here, the Danube still exerts its functions associated with each inundation, such as:

. erosion effects within the alluvial plain with a possible formation of evorsion lakes, ie. lakes and pools created by a strong turbulent water flow
. flushing of the water system of the alluvial plain, inhibiting the silting of most water bodies forming this system (on sites with a rapid water flow)
. sedimentation of suspended matter of the Danube and along with it fertilization of the alluvial plain with nutrients such as

D. F. Whigham et al. (eds.), Wetland Ecology and Management: Case Studies, 127–130.

N and P (on sites with a slow water flow, ie. mainly above the average watermark of the alluvial water bodies

. prevention of the growth of the macrophytic vegetation in those channels of the alluvial plain which are exposed to frequent flushing

. removal of large quantities of aquatic organisms and dramatic changes in the physical and chemical parameters of the water.

On the other hand, a long-term low level of the Danube may result in the alluvial water system drying out with all its consequences such as high decomposition rate of organic matter. In the alluvial plain up-stream from Altenwörth most of these processes can no longer be observed, or they are at least considerably reduced. The Danube is allowed to inundate only from a flow rate of 5,800 m^3s^{-1} upwards, a flow rate which has been rare since the power station was opened in 1977. But even then the amount of inundating of the alluvial plain meets the requirements for its proper ecological functioning on extremely rare conditions so that the original condition of the water system cannot be sustained. The purpose of this paper is to compare the two sections of the Danube, Hainburg and Altenwörth.

COMPARATIVE DESCRIPTION OF THE DANUBE ALLUVIAL PLAINS NEAR HAINBURG AND ALTENWÖRTH.

1. Sedimentation: Within the flood zone of the alluvial plain near Hainburg, the sediment accumulation is generally small. Bodies of water exposed to frequent flushing caused by inundation by the Danube show no significant sedimentation; in other bodies of water the sediment depth surpasses 50 cm only at six sites of a total of 32 sites investigated. The accumulated, predominantly inorganic, sediment exhibited traces of sapropelization only in a few cases. In contrast to this, bodies of water north of the Danube near Altenwörth have become partly dry while in those still existing 50% of the 20 sites inspected have a sediment layer deeper than 50 cm and at three sites its depth exceeds 100 cm. Even in the channels which, before the construction of the water works, belonged to the flooded portion of the river Kamp, accumulation of sediment in the order of 10 cm was recorded. This corresponds to a sedimentation rate of more than 1 cm yr^{-1}. Moreover, the sediment profiles of most of the aquatic sites exhibited sapropelic layers; in dry or almost dry water bodies , on the other hand, the sediments showed signs of terrestrial soil formation. This indicates that, apart from infrequent flooding, the fluctuations of groundwater levels are reduced and no longer provide conditions for the formation of subhydric soils.

2. Chemical properties: Chemical data from both sites are listed in Table 1. No inundation occurred in either section of the alluvial plain during the time of our investigations in 1983 and 1984. In general, sampling stations near Altenwörth were less acid. More sampling stations near Hainburg generally had a higher conductivity than the main stream of the Danube, obviously caused by the groundwater which is enriched with ions from the local soil and subsoil. In contrast to this, the river Kamp, which drains the igneous rock area of the "Waldviertel", exerts its influence on most of the existing bodies of water north of the Danube near Altenwörth, and therefore lower conductivity values than in the river Danube itself are typical of the alluvial water system there. Higher conductivities have been found only in artificial gravel pits and the Krems River, which serve as discharge areas to the river.

Table 1. Summary of water chemistry data for two areas along the Danube. The data are frequencies for 15 sites sampled in each area. Values with an * are based on a sample size of 14 rather than 15.

	Altenwörth	Hainburg
pH		
6.5-7.5	33.3	71.5
7.5-7.8	20.0	21.4
7.8-9.2	46.7	7.1
Conductivity (μS)		
200-300	60.0	26.7
301-400	40.0	20.0
401-500	0.0	20.0
501-700	0.0	33.3
Total Phosphorus ($\mu g\ l^{-1}$)		
15-100	33.3	50.0
101-150	20.0	28.6
151-320	40.0	14.3
321-600	6.7	7.1
Total Nitrogen ($\mu g\ l^{-1}$)		
50-100	0.0	7.1
101-200	6.7	0.0
201-400	13.3	0.0
401-700	60.0	28.6
701-1400	20.0	64.3

Figure 2. Photographs of wetland areas near Stapfenreuth along the Danube. The upper photograph is a typical aspect of the Rosskoffarm near Stapfenreuth with Phragmites in the background and Nuphar in the foreground. The bottom photograph shows an area dominated by Salix alba.

Total phosphorus and nitrogen concentrations were similar in both sections and they can exceed 500μg l^{-1} and 2,000μg l^{-1}, respectively. Concentrations exceeding 2,000μg l^{-1} of NO_3 have, however, been found only in the water system influenced by the River Kamp near Altenwörth.

3. Macrophytic vegetation and algae: Typical wetland habitats are shown in Fig. 2. About 20 common macrophyte species were found in waterbodies of the alluvial plain near Altenwörth in 1983. Approximately half occurred in man-made gravel pits (Kusel oral comm.). In contrast, almost 60 species occurred near Hainburg. Several, such as Potamogeton acutifolius, Ranunculus rionii, Veronica catenata, and Stratiotes alioides are rare in Austria and Central Europe. The greater diversity is undoubtedly related to the presence of a greater diversity of wetland habitats in the unaltered areas.

As to the algae, the waters of both sites are predominated by bluegreens, such as Anabaena flos-aquae, flagellates (Eudorina, Pandorina), dinoflagellates (Peridinium) and diatoms (Synedra acus, Fragilaria crotonesis, Asterionella). One of the latter, Melosira binderana, is typical of the Danube and occurs near Hainburg in bodies of water frequently flooded and near Altenwörth only in the former river bed of the Danube (Kusel oral comm.).

4. Zooplankton and benthic invertebrates: Differences in zooplankton of both sites can be defined. There are, however, indications that the diversity of the site near Hainburg is much greater than near Altenwörth.

With respect to the benthic fauna there is no doubt that its variety in the Hainburg area exceeds that at Altenwörth. Besides the crustaceans typical of macrophyte beds of the Danube, such as Corophium sp. and Limnomysis benedeni, which increasingly invade the upper part of the river because of the construction of dams and power plants, a wealth of rare species occurs in the waters near Hainburg.

The site near Altenwörth, in contrast, is inhabited by common organisms. Among crustaceans, Dunhevedia crassa and Physocypria fadeewi, so far not known from Austria, may be mentioned, and among other aquatic invertebrates (e.g. 27 species of Odonata, 55 species of Coleoptera) about 40 species are endangered in Austria, or even in the whole of Central Europe. Thus the dragon-fly Aeschna viridis needs Stratiotes alioides for its reproduction and because of the rarity of this plant, it has a very limited distribution in Austria. On the other hand, the unique diversity of molluscs (35 species) in the backwaters near Hainburg, which is not surpassed by any other site in Austria, offers valuable food resources to fish and spawning facilities to the Bitterling (Rhodeus sericeus). In contrast to the area near Hainburg, the waters of the alluvial plain of Altenwörth have a mollusc fauna comprising only 16 common species. Analysis of cores will provide further information about the mollusc diversity before the construction of the power plant.

An even greater discrepancy between the backwaters of Hainburg and Altenwörth has been found with respect to the spawning places of amphibians, mainly frogs of the genera Rana, Bombina and Bufo. The number of such sites is more than ten times greater in the area of Hainburg than near Altenwörth. This is partly due to the re-shaping of old backwaters and new gravel pits with steep banks: such sites are not easily accessible to these animals.

Similarly, the differences between the fish faunas at the two sites is obvious (Merwald 1981). In contrast to the waters of Hainburg, where thirty fish species were observed, only four occurred near Altenwörth at the time of our investigation. The only exception is the former river bed of the Danube which, at the time of our investigation, was still connected down-stream with the river and not yet influenced by any power plant. About 15 species occurred here. Since then a new power plant has been opened, north of Wien (Greifenstein), and it will be of interest to learn about the changes after this most recent impact. The reasons for the poverty of ichthyofauna in the other backwaters is obviously connected with

- lack of benthic zones with gravel
- poor development and low species variety of macrophytes
- separation of the majority of backwaters from the Danube.

CONCLUSIONS

Summarizing the results of this comparison, it seems necessary to emphasize the lack of proper technologies connected with the construction of power plants in open plains, which would correspond to the needs of the alluvial aquatic ecosystems. Obviously, the complete sealing of the river, which prevents groundwater fluctuations in the alluvial palin and thus the flushing of the gravel layer form below, together with the absence of inundations, results in an ecological impoverishment of backwaters. Therefore, it seems irresponsible to expose the unique riverine area of the alluvial plain between Wien and Hainburg to such an impact, since it represents the last large area of a riverine landscape in Europe.

REFERENCES

Löffler, H. et al., 1976: Limnologische Untersuchungen am Eberschnuttwasser-Gutachten im Auftrag der Stadt Wien, Austria. 80 p.

Löffler, H. et al., 1983: Limnologische Untersuchung zur Standortfrage de Donaukraftwertkes Hainburg/Deutsch Altenburg - Gutachten im Auftrag des Bundesministeriums für Land und Forstwirtschaft. 110 p.

Merwald, F. 1981: Die Veranderung der Fish-fauna eines Donau-Augrabens in fünfzig Jahren. ÖKO-L 3/1: 19-23

VEGETATIONAL CHANGES IN RESPONSE TO DRAINAGE AT SAROBETSU MIRE, N. HOKKAIDO, JAPAN

Koji Ito and Leslaw Wolejko
Graduate School of Environmental Science,
Hokkaido University,
Sapporo 060, JAPAN

ABSTRACT

Oligotrophic wetlands in Hokkaido are still sufficiently preserved in the mountainous areas. However, rare elements of mesotrophic fens and bogs in lowlands are on the verge of extermination. Thus, necessary efforts in nature protection should be oriented towards preservation of wetlands and species in these particular types of mire systems.

In his concluding remarks on the future of mires Goodall (1983) stated, that although the existence of mires as an ecosystem is not endangered, there are areas where the destruction of these wetlands has been almost completed. The most vulnerable are mires in areas of high population density and also those formed on the border of their geographical, climatically controlled distribution (Terasmae 1977).

In Japan, both of those situations have resulted in deterioration of mires almost everywhere on the main islands of Honshu, Shikoku and Kyushu. In the peat forming lowlands of Hokkaido, the transformation and destruction of mires is a relatively new phenomenon, dating back to the beginning of this century, but has achieved considerable scope in the sixties. During that time a large wetland conversion scheme supported by the government commenced and more than 50% of the lowland mires (originally 200,642 ha, Yano et al. 1980) have been converted into agricultural and urban areas (Fig. 1).

There are still a considerable number of small mountain mires, both in Hokkaido and northern Honshu, which are relatively safe from total destruction. In many cases, however, the interesting and rare flora and vegetation cover are affected by treading by numerous hikers and 'nature loving' visitors (Tachibana 1976; Ito 1983).

The process of mire transformation has been best studied in Hokkaido where the amount of conversion varies considerably depending on climatic conditions, accessibility, and distance from development centres. Natural mire vegetation has already almost disappeared from the southern, western and northwestern parts of the island but there are still some extensive and unchanged mires in the eastern region, namely the Kushiro, Furen and Kiritappu Mires. The reason for the spatial variation in mire transformation is well understood when the cold climate of the eastern coast, influenced by cold Kurile Current, is compared with the milder climate of southern and western Hokkaido (Kojima 1979). The latter, improved by the warm Tsushima Current, allows rice cultivation on the developed surfaces of former mires. An example of this is the Ishikari Plain, where about 55,000 ha of peatlands were transformed into paddy fields.

An evaluation of the relative importance of different impact types on mires in Hokkaido is given in Table 1. The biggest threat is posed by agriculture and different forms of infrastructure development on the lowland mires. For mountain mires, direct human impact is the most important. An example of the transformation process for a lowland mire in the northwestern part of Hokkaido is presented in Fig. 2. During a course of time the progress in agricultural development, channel digging, road construction and river regulation can be seen.

Changes in the mire ecosystems can be expressed at three levels (landscape, plant associations and species). Transformation at the landscape level is most conspicuous when large areas of fen, bog, and swamp forest are substituted by uniform and regular human-made fields and pastures (Fig. 2).

D. F. Whigham et al. (eds.), Wetland Ecology and Management: Case Studies, 131–134.

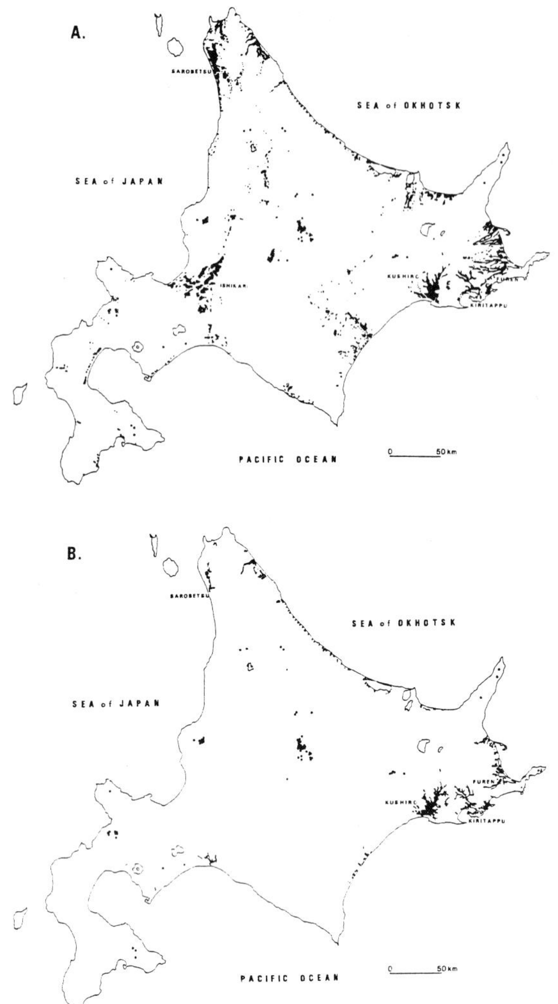

Figure 1. Past and present distribution of mire vegetation in Hokkaido; A. original distribution (based on the map by Sakaguchi 1979); B. present distribution (based on the map by Ito et al., 1982). Circles symbolize small mountain mires.

At the level of plant associations, changes can also be detected. In the Sarobetsu Mire the vegetation has been invaded by other species as a result of mire conversion. The dominant species, representative of a dwarf-bamboo genus Sasa, gives the very characteristic physiognomy to the areas affected by hydrological disturbances. Channel digging is the most influential factor, allowing the advancement of Sasa community along newlybuilt watercourses. The replacement of a poor-fen community dominated by Moliniopsidetum japonicae by Sasa palmata (Bean) Nakai facies is the most remarkable feature of these areas. In addition to changes in dominance, there is a decrease in species diversity and almost complete disappearance of character species on Oxycocco Caricentum middendorfii and Sphagnetum papillosi associations. Some species characteristic of bog vegetation (Vaccinium oxycoccus L., Chamaedaphne calyculata (L) Moench., Drosera rotundifolia L., Sphagnum papillosum Lindb., etc.) have also been reported to decline (Tachibana and Ito 1980).

The decline in the floral diversity of mires has also been documented (cf. Jasnowski 1977; Mannema et al. 1980; see also Kornas 1983, for further references). The main cause is usually disappearance or transformation of the suitable environmental conditions for mire plant growth.

The indigenous flora of the Sarobetsu Mire is characterized by numerous rare bog plants, and by relative richness of orchids. From the 71 species of Orchidaceae native for Hokkaido (Tatweaki 1954), several were reported from the mire. Their survival, as well as that of other bog plants, is highly endangered by drainage. This applies to such species of boggy orchids as Amitostigma kinoshitae (Makino) Schltr., Habenaria yezoensis Hara var. longicalcarata Miyabe et Tatew., Platanthera tipuloides Lindl. var. nipponica (Makino) Ohwi, Pogonia japonica Reichenb. f. and Eleorchis japonica (A. Gray) F. Maek., among which the latter two species are in Sarobetsu Mire at their northern distributional border. The list of other endangered bog species contains Drosera anglica Huds., Scheuchzeria palustris L. and Lycopodium inundatum L.

SUMMARY

Changes in the Sarobetsu mire, Northern Hokkaido, Japan can be expressed at the following three levels.

(1) The transformation of the landscape has been most prominent in the appearance of uniform and regular human-made fields and pastures in place of mire landscape.

(2) At the level of plant associations, the invasion of Sasa into mire vegetation and the replacement of the Moniliopsidetum by Sasa facies are the most predominent.

(3) Noteworthy changes at the floral level have been the decrease of endangered bog species and orchids.

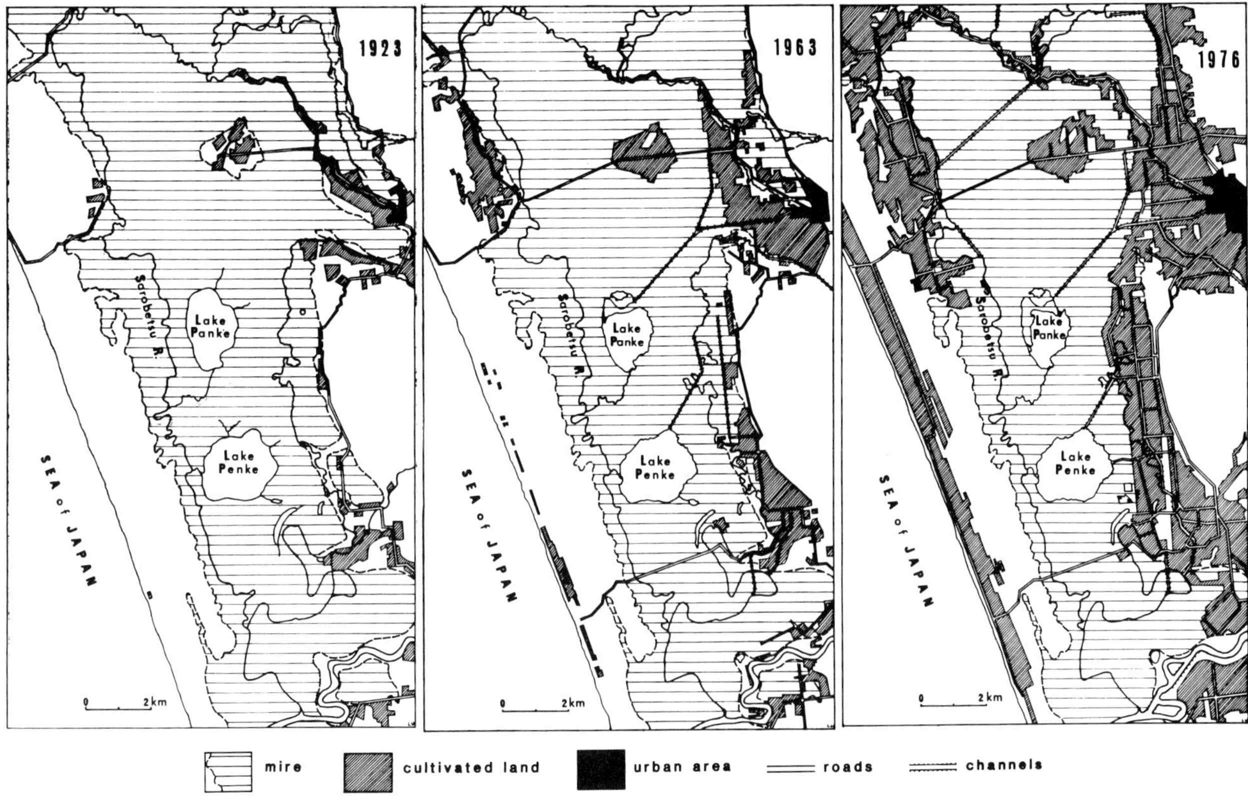

Figure 2. Transformation of the Sarobetsu Mire, Northern Hokkaido, 1923, 1963 and 1976.

Table 1. Relative importance of different human impact types on mire ecosystem in Hokkaido.

Type of Impact	Description	Relative importance in Hokkaido
A. Direct		
1. Change of function		
a. urban development	Occupation of mire space for construction, housing, dump storage, sports facilities, etc.	XX
b. agricultural	Agricultural crop production after drainage, soil dressing, etc. Utilization for hay and pasture.	XXX
2. Peat cutting	Excavation and removal of peat.	X
3. Burning	Intentional or incidental fire of mire surface.	XX
4. Direct human impact	Trampling by visitors, plant collection.	XXX
B. Indirect		
5. Disturbances in hydrological regime	Change in quantity of water, fluctuations of water level etc., induced by e.g. river short-cutting and embankment, construction of roads and railways.	XXX
6. Change of water quality and nutrient upply	Eutrophication (leaching from agriculture, industrial and urban pollution). Acidification.	XX

XXXX - very strong; XXX - strong; XX - moderate; X - negligible.

REFERENCES

Goodall, D.W. 1983. Conclusion - the future of mires. pp. 395-396. In Gore, A.J.P., ed. Mires: swamp, bog, fen and moor. Ecosystems of the World 4B, Elsevier, New York, NY. USA.

Ito, K. 1983. Man's impact on the wetlands in Japan. pp. 327-334. In Holzner, W., Werger, M.J.A. and Ikusima, I., eds. Man's impact on vegetation,. Dr. W. Junk Publisher, The Hague, The Netherlands.

Ito, K., Shimizu, M. and Koga, M. 1982. Ecosystems of Hokkaido. Map in scale 1:600,000 based on LANDSAT images. Japan Foundation for Shipbuilding Advancement. Tokyo, Japan.

Jasnowski, M. 1977. Probleme und Methoden des Moorschutzes in Polen. Telma 7: 215-239.

Kojima, S. 1979. Biogeoclimatic zones of Hokkaido Island, Japan. Journal College Liberal Arts, Toyama University 12: 97-141.

Kornas, J. 1983. Man's impact upon the flora and vegetation in Central Europe. pp. 277-286. In Holzner, W., Werger, M.J.A. and Ikusima, I., eds. Man's impact on vegetation, Dr. W. Junk Publisher, The Hague, The Netherlands.

Mannema, J., Quente-Boterenbrood, A.J., Plate, C.L., eds. 1980. Atlas of the Netherlands flora. I. Extinct and very rare species. Dr. W. Junk Publisher, The Hague, The Netherlands. 226 p.

Sakaguchi, Y. 1979. Distribution and genesis of Japanese peatlands. Bulletin Department Geography Tokyo University 11:17-42.

Tachibana, H. 1976. Changes and revegetation in Sphagnum moors destroyed by human treading. Ecological Review 18:133-210.

Tachibana, H. and Ito, K. 1980. Phytosociological studies of the Sarobetsu Mire in the northern part of Hokkaido, Japan. Environmental Science Hokkaido 3:73-134.

Tatewaki, M. 1954. Phytogeographical studies on Orchidaceae in the islands of the North Pacific. Acta Horti Gotoburg. 19:51-112.

Terasmae, J. 1977. Postglacial history of Canadian muskeg. pp. 9-30. In N.W. Radforth and Brawner, C.O., eds. Muskeg and the northern environment in Canada, University of Toronto Press, Toronto, Canada.

Yano, Y., Akazawa, T. and Umeda, Y. 1980. Present status and problems for agricultural use of peatland in Hokkaido, Northern Japan. pp. 501-505. In Proceedings 6th International Peat Congress Duluth, MN, USA. 1980. Society of International Congress, Duluth, MN. USA.

THE APIARY OF THE MANGROVES

K. Krishnamurthy
Centre of Advanced Study in Marine Biology
Annemalai University, Parangipettai 608 502
Tamil Nadu, India

ABSTRACT

Mangroves provide a variety of resources that can be used by man. For centuries, people in India have been gathering honey made from the nectar of mangrove species. Honey extraction, to date, has been a cottage industry but there is a large potential for development of a larger industry that would not have a negative impact on mangrove systems. In this paper I describe the nature of the honey industry in Indian mangroves.

INTRODUCTION

Mangrove forests are essentially tropical and subtropical. They typically occur between 23 $1/2^{0}$N and 23 $1/2^{0}$S of the equator, but occasionally may also extend beyond this limit. The mangrove distribution may extend up to the warm temperate area, though in diminished numbers as in Japan (Kagoshima, Okinawa) and New Zealand (Auckland). Mangroves provide many benefits to human societies. In this paper I describe how mangroves are used for the gathering and/or husbandry of honey.

DESCRIPTION OF THE STUDY AREA

The Indo-Pacific Region was the ancestral home or cradle of evolution of the mangrove vegetation. There are two genera of the mangrove vegetation originally restricted to the Old World only - viz., Rhizophora and Avicennia. They, later on, spread to the New World through the oceanic currents which wash their coastlines (Chapman, 1975) and through the maiden voyages of discovery in various parts of the world undertaken by many sailors to colonise new lands over the centuries.

In the Indo-West Pacific Region, their greatest diversity and maximum luxuriance are observed. (Macnae, 1968). The reasons for this may be the climatic stability, optimum physiographic conditions like soil texture, sediment composition, formation of many great deltaic regions of the mighty rivers, congenial atmospheric and water temperature, proper admixture of seawater and freshwater.

Thus, this region offers the ideal growth conditions for this vegetation which result in its maximum luxuriance.

There are about one hundred species of mangroves which include also their main associates. In India there are about sixty species of the mangroves and associates. They are distributed unevenly along the Indian coastline. The mangroves of India and the world can be classified under three categories:

1. Deltaic mangroves
2. Estuarine-backwater mangroves
3. Insular mangroves

1. Deltaic mangroves are found on the East Coast of India. They are situated in the tail-end delta portions of the mighty rivers such as the Ganga (the Ganges), Brahmaputra, Mahanadi, Godavari, Drishna and Cauvery. These Deltaic mangroves, the most luxuriant in India, are highly diverse and are also characteristic of the deltaic portions of all major rivers of the Indo-Pacific such as the Brahmaputra, Irrawaddy, Medong, etc.,

2. Estuarine and backwater mangroves are found on the Western Coast of India where there are no major rivers except the Narmada and Tapti; which do not form deltas but true estuarine mouths.

3. Insular mangroves are found on the islands of India, particularly on Andaman and Nicobar. They are much better preserved than any other Indian mangroves.

The three types of mangrove forests in India cover an area of about 6,670 sq. mi. (Krishnamurthy, 1986). The Eastern Coast accounts for 70%, the Western Coast for 13% and the insular mangroves (Andaman and Nicobar) for 17%.

D. F. Whigham et al. (eds.), Wetland Ecology and Management: Case Studies, 135–140.

MANGROVE FLORA THAT ARE OF VALUE FOR HONEY PRODUCTION

The inflorescence of mangroves produces some of the best known honey. The mangrove species which account for major honey production are: Excoecaria agallocha, Phoenix spp. Rhizophora mucronata, Avicennia marina, A. officinalis, Ceriops tagal, Ceriops decandra and Bruguiera gymnorrhiza (Fig. 1). The least preferred species is Aegiceros corniculatum. Another mangrove species, Heriteria fomes, favoured by the bees, is a freshwater species and the diminished flow of freshwater in the region has led to its total disappearance. This species is known as Sundari in the Bengali language and is the origin of the name "Sunderbans."

In the Sunderbans area (which accounts for about 90% of honey production among the mangroves of India), the mixed natural formations of Phoenix - Excoecaria are favoured most by the bees. Some honey is also collected from the mangroves of the Mahanadi - Godavari - Drishna area and from the mangroves of Andaman and Nicobar. The main species of bee is Apis dorsata. Licenses are issued to collect honey in the Sunderbans. The extraction technique used to collect honey may be primitive. Significant quantities of honey are produced from the Sunderban mangroves of India, Bangladesh and the Southwest Florida mangroves of the U.S.A. Since the Sunderbans is spread between India and Bangladesh, forming one continuous block, the species composition of the mangrove vegetation and the method of honey collection are very much the same. The production and trade of honey in the region is of economic importance to the communities of the peoples of the Sunderbans of India and Bangladesh.

The honey production from the different species belonging to mangrove vegetation is described by Chakrabarti and Chaudhuri (1972). The essential information, given below, is based mostly on their work.

Excoecaria trees were observed to possess honeycombs of native bees more than any other vegetation (Fig. 2). However, this tree by its size and shape does not form an ideal choice. The mixed Phoenix-Exocoecaria formation would appear to form the ideal choice. The cool atmosphere and the moist substratum under the shade of this tree and the Excoecaria branches were found to contain the maximum number of combs per unit area. Xylocarpus species are avoided. Sonneratia apetala, the tallest tree of the Sunderbans, does not contain any appreciable number of honey combs. The Ceriops species account for 11% of the combs noticed. Most honey is collected in mid-April although collecting is done over a three month period (Fig. 3).

AMOUNT OF HONEY HARVESTED AND ITS ECONOMIC IMPORTANCE

The amount of honey collection from the 955 km^2 buffer areas of the Tiger Reserves of the Indian Sunderbans has more than doubled during the last five years. The average annual yield before the launching of the Project Tiger

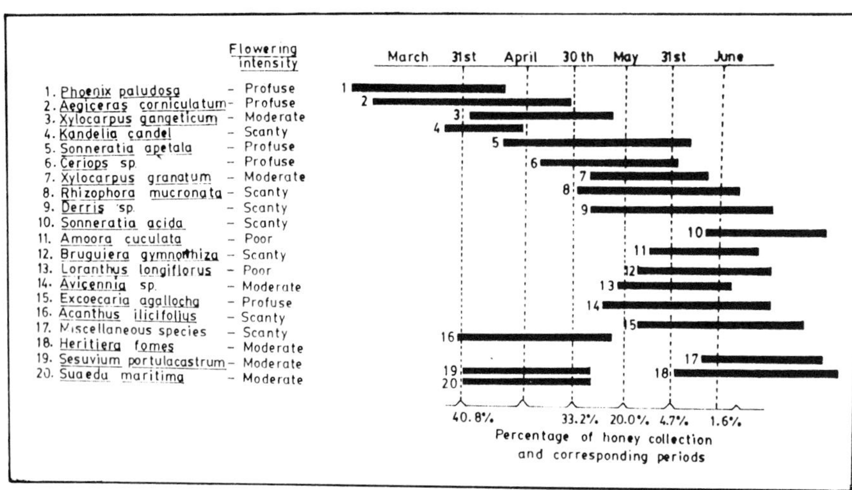

Figure 1. Major honey-producing mangrove species and their period of availability (black bars) for honey collection.

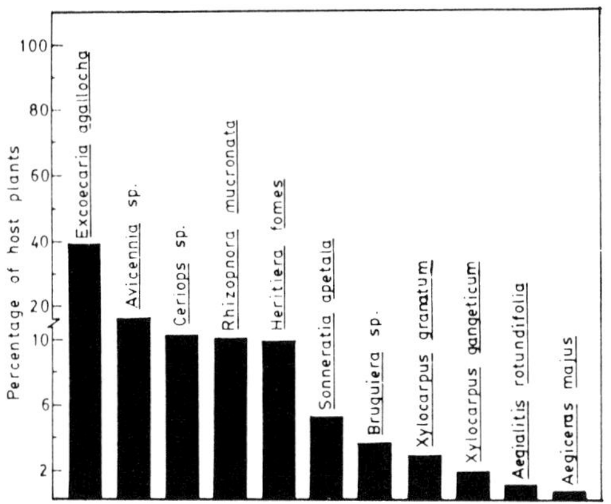

Figure 2. Relationship between occurrence of native honeycombs and mangrove plant species.

(World Wildlife Fund is also involved in funding this) in that zone (later to become a buffer area) was about 400 quintals (1 quintal = 100 kg). Now, during the last 3 years, the yields per annum from the buffer areas were as follows:

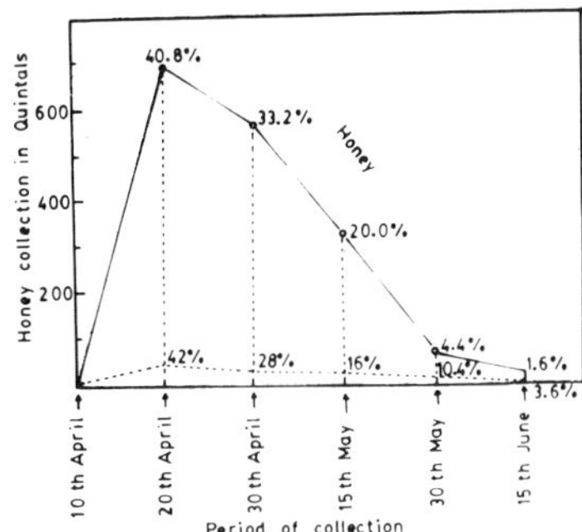

Figure 3. Amount of honey in quintals (1 quintal = 100 kg) and percent honey collected during the April 10 - June 15 collecting season.

Year	Yield of Honey in Quintals
1983	510
1984	615
1985	775

The reason for the markedly improved yield of honey is ascribed to protection, preservation and conservation of the flowering plants and trees. So more nectar is available for the bees to gather.

Honey is an essential ingredient in the traditional Indian system of medicine called "Ayurveda" (the science of long life). Honey is applied over burns and is used as a good preservative. Table 1 gives information on honey collection for a decade.

Table 1. Information on ten continuous years of production of honey, wax, etc. (in quintals) collected from the Sunderbans mangroves of India (after Chakrabarti and Chaudhuri, 1972).

Year	Honey	Wax
1963 - 64	1323.8	100.1
1964 - 65	599.6	34.2
1965 - 66	959.2	67.8
1966 - 67	773.0	62.6
1967 - 68	889.5	128.5
1968 - 69	1718.6	71.0
1969 - 70	680.5	57.3
1970 - 71	912.9	74.7
1971 - 72	890.0	72.1

METHODS OF HONEY COLLECTION

There seem to be three major methods used in honey collection. They are practised, although the methodology is still primitive, in the Indo-Pacific countries - of India and Bangladesh and Africa. They are broadly divisible into:

1. Bee hunting or raiding
2. a. Apiculture or bee maintenance
   b. Bee keeping

1. In the bee hunting or raiding method, the entire colony is destroyed, as it is set on fire and depending on lunar phases in periodicity - when the bees are expected to store and fill the honey comb to their maximum capacity. The ember that is placed underneath the comb, as it is fanned, drives away the bees. The hunters make no effort to conserve the bee population. This procedure is often difficult as most bee hives are located more than 5 meters above the ground (Fig. 4).

In certain areas of Africa and India straw containers or clay pots are hung in trees to attract the wild bees or induce colony formation. After sufficient honey is accumulated, the bees are killed and honey collected. In this method, time is saved which

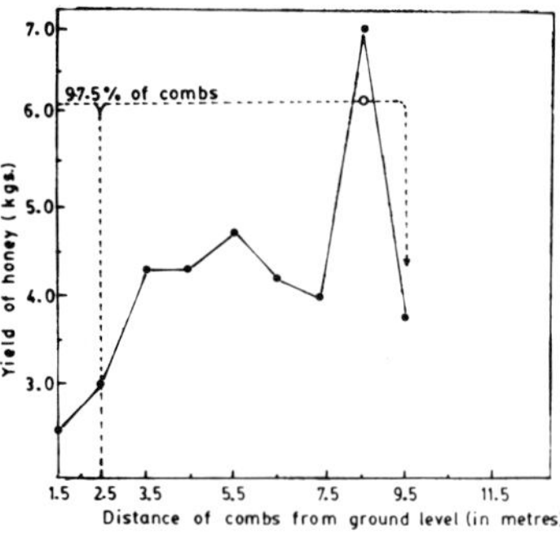

Figure 4. Relationship between honey yield (kg) and distance in meters of honeycomb above ground.

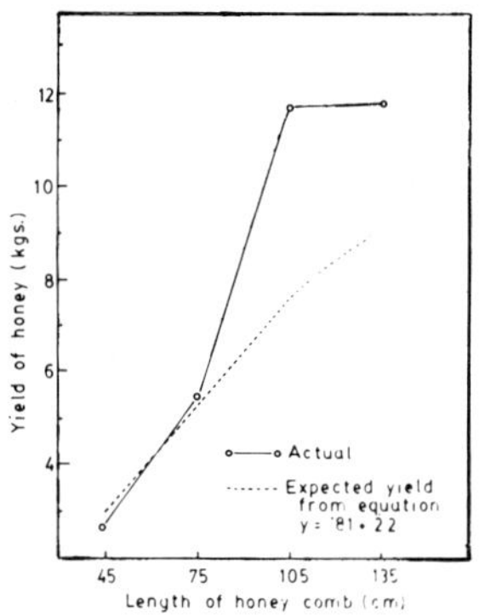

Figure 5. Relationship between honeycomb size (as length in cm) and honey yield (kg).

would be otherwise wasted in locating the wild bees or their hives.

2a. The apiculture technique of bee maintenance represents the best managed relationship between bees and man. The bee colony is provided shelter and habitation, usually made of plant materials, gourds, clay pots, etc. The bees attach honey and brood combs to the inside of containers. The honey is collected from the honey combs leaving free the broodcombs.

In this method both wax and honey arc collected perennially forming an ideal renewable source of honey supply and perennial source of income. The comb is not destroyed.

2b. The application of advanced techniques involves a knowledge of the behaviour of bees. The method of beekeeping involves applicaton of intermediate and high technology, as in Florida. Thus manipulation of a bee colony is possible for enhancement of honey production. In the intermediate level movable comb hives are used.

FACTORS AFFECTING THIS COTTAGE INDUSTRY

The following are the factors that affect this cottage industry:

The Species of the Honey Bee

Honey production is principally carried out by: a) Apis dorsata (rock bee), Apis florea (little bee), Apis indica (Indian bee), and Apis mellifera (dammer bee).

Apis dorsata is the dominant species found in the wild in the Sunderbans. They live also in the nearby hills and forests.

Climatological Conditions

It is interesting to note that Apis dorsata migrates to the Sunderbans between March and July. They cover hundreds of miles to reach the area during the season when atmospheric humidity is between 75% and 85%. During this season the peak bloom period of the mangrove vegetation also occurs. The honey bee does not like the rainy season for honey production. Excess rain affects the yield of honey. In the same manner a drought period is not conducive for honey production.

Size of the Comb

Beehives are usually big, measuring up to one meter in length. Trees usually contain only one beehive although two hives occurred on the same tree in between 5% and 10% of the trees sampled.

Honey yield varies with comb size. Combs of 0.03 m^3 volume yield about 8 kg of honey (Fig. 5). and combs of 0.05 m^3 yield about 14 kg of honey. Combs larger than 0.05 m^3 do not generally occur (Chakrabarti and Chaudhuri, 1972).

Relationship to Vegetation:

The amount of honey produced in an area is primarily dependent on the vegetation. Nectar collection by the bee is directly proportional to the number of species in the vegetation. In the Sunderbans there are about sixty species of mangroves and their associated vegetation. Hence a wide 'variety' of nectar is gathered by the wild bees. Combs are found on many different species. The duration of the flowering season, which determines the availability of pollen and nectar, is also longer in area with high species diversity.

Figures 1-4 show the pattern of honey production. The angle of exposure of the honeycomb to the sun will also determine the quantity of honey it contains. Combs that are situated at right angles to the sun have a high honey content. Combs accumulate honey on their lower portions. Fresh combs are constructed on new sites.

Choosing the Optimum Period of Production

The flowering of the mangrove vegetation is at its peak usually during April to July. Humans can do little to influence the duration of the flowering season. A good and healthy colony maintenance during the peak season of bloom is a sign of good management.

The Problem of Pests

The natural pest problem is not very serious. Crabs like Scylla serrata are found on the beehives as are monkeys, bears, etc. The monkeys usually smear their body with mud as an antidote to the bee sting before venturing on the combs. Bears are also known to relish honey and "hunt" for the honey. The tigers in the area are known to show occasionally their preference for honey.

Presence of Salt

The presence of salt in the atmosphere and upon the vegetation as crystals (by salt excreting plants) and in the water of the area affects the health of the bee. Even a low concentration of 0.125% salt in the diet of the bees induces mass mortality (Dadant and Sons, 1982). Thus, the health of the bees and maintenance of the beehives will suffer a setback if drought conditions are prolonged or frequently recur. Normally this does not happen as monsoonal rainfall alternates with the summer season. The decline in the freshwater sources of supply also poses a serious problem. The bees need freshwater for their good health and for honey production. They use the freshwater to mix with the honey they produce before use. Freshwater also helps to cool the beehives. In hot climates droplets of water are lodged around the comb to produce evaporative cooling. The bees fan their wings to enhance the cooling effect. In the mangroves more honeycombs are found in areas adjacent to freshwater.

CONCLUSIONS

Honey extraction is a traditional activity. Honey harvest from the mangrove forests is not a widespread activity, but is traditionally restricted to certain mangrove areas only. Moreover it is confined to a handful of countries - out of about 100 countries where we have mangroves. This is surprising indeed as the mangrove forests have a high potential for honey production. The rather poor nature of honey collection from these areas, is due to:

1. The rather inaccessible situation and remoteness of the mangrove localities.
2. The difficult mangrove terrain.
3. The fear of wildlife.
4. It is easier to collect honey from other forests than from the mangrove source.
5. The mangrove areas are also situated in the cyclone - prone zones (like the Bay of Bengal) and also flood - prone; periodic drought would also drive away the bees.

The decline in freshwater flow in the Indian Sunderbans has resulted in extinction of freshwater - loving mangroves like Lypa fruticans and Heriteria fomes and Phoenix spp. etc. This decline is also posing problems for the denizens, including bees.

The rural folk in general are illiterate. They need to be "educated" in the modern techniques. Honey is a good source of non-perishable and nutritious food. It is a profitable venture. Bee-keeping is a labour intensive cottage industry. It fits well with the concept of agro-development also. It can be practised by the entire family. It is an employment generating industry which can be integrated with technology transfer in modern apiary practices and rural development. It merges with ecologically compatible development and forestry management.

ACKNOWLEDGEMENTS

My thanks are due to the authorities of Annamalai University for facilities provided and to the Project Tiger authorities of the Indian Sunderbans for their courtesy, hospitality and for useful discussions.

REFERENCES

Chakrabarti, K. and A.B. Chaudhuri, 1972. Wildlife biology of the Sunderbans forests: honey production and the behaviour pattern of the honey bee. Science and Culture 38:269-276.

Chapman, V.J. 1975. Mangrove biogeography pp. 3-22. In Proceedings of a International Symposium on Biological Management of Mangroves Vol 1. eds. Walsh, G. E., Snedaker, S. and Teas, H. J. , eds. Gainesville. Univ. Florida, USA.

Chaudhuri, A.B. and K. Chakrabarti. 1973. Wildlife biology of the Sunderbans forests: A study of the habit and habitat of the tigers. Bulletin of the Botanical Society of Bengal 26:63-66.

Dadant and Sons eds. 1982. The hive and the honey bee: Dadant and Sons, Inc. Hamilton, IL, USA.

Honey production techniques in mangrove environments. 1984. pp. 30-37. In Handbook for mangrove area management (eds.) Hamilton, L. S. and Snedaker, S. C., eds. The East-West Centre. Honolulu, HI, USA.

Krishnamurthy, K. 1986. The changing landscape of the Indian mangroves: pp. 119-126. In The Mangroves ed. Bhosale, L. J., ed. Shivaju Univ. Press, Kochapur, India.

Macnae, W. 1968. A general account of the fauna and flora of mangrove swamps and forests in the Indo-West Pacific Region. pp. 73-270. In Advances in marine biology Vol. 6. Russell, F. S. and Yonge, C. M., eds. Academic Press, London, U.K.

CHARACTERISTICS OF A SEASONALLY FLOODED FRESHWATER SYSTEM IN MONSOONAL AUSTRALIA

C. Max Finlayson[1], Ian D.Cowie[2] and Bruce J. Bailey[3]
Office of the Supervising Scientist
Alligator Rivers Region Research Institute,
Post Office,
Jabiru. N.T. 0886
Australia

ABSTRACT

Magela Creek is located in the monsoonal area of northern Australia. This paper describes the physico-chemical and biological characteristics of the 250 km^2 of the Magela floodplain.

During the wet season the floodplain is covered with water that is neutral to acidic, very soft and has low ionic concentrations. The velocity of the flood-front and the duration of the dry season greatly affect the physico-chemical status of the water. Surface water temperatures can reach 37° and the general stratification pattern is classical polymixis.

The phytoplankton community is very diverse with more than 160 diatom taxa and 530 taxa from other groups. Chlorophyll values were not generally high except during blooms. The macrophytic flora was also diverse with 225 species, 139 of them being annuals. The floristic composition and foliage cover of the 10 main macrophyte communities varied seasonally; the abundance and floristic richness of the wet season contrasting with the sparse cover of the dry season.

The vertebrate and invertebrate flora were also diverse. The invertebrates demonstrated vast temporal and spatial differences in species abundance. Species diversity was greatest during the wet season. Similarly, the distribution and habitat selection and abundance of the fishes and amphibia were influenced by the occurrence of floodwaters during the wet season. The water birds were also numerous and diverse. Their usage of the floodplain varied seasonally with about 200,000 occurring during the dry season when food resources were scarce elsewhere.

Stimulation for this documentation came from proposals to discharge excess water from a uranium mine to Magela Creek. The biological documentation was one step in the process of determining standards and protocols to govern any releases of water. The assessment of any impact from the mining operation on the floodplain was complicated by changes brought about by other activities in the area. Control of feral animals and occurrence of aggressive alien plant species are two such activities.

1 1, WRB
Slimbridge Alos.
AL 2 7BX
England

2 CONT
PO Box 469
Palmerston N.T. 0831
Australia

3 Soil Conservation Service of
New South Wales
PO Box 10
Wagga Wagga N.S.W. 2650
Australia

D. F. Whigham et al. (eds.), Wetland Ecology and Management: Case Studies, 141–162.

INTRODUCTION

Interest in the Magela Creek ecosystem of northern Australia has increased following the discovery of two large uranium deposits - Ranger and Jabiluka - in the catchment The deposit at Ranger is currently being mined and milled at Jabiru (see Fig. 1 for all localities). Government approval for these activities was largely dependent on conditions and guidelines established by the Ranger Uranium Environmental Inquiry (Fox et al. 1977) and later developed into a series of environmental requirements.

The inquiry suggested that the biota of the ecosystem could be at risk from the mining operations, if an excessive amount of waste water were to be released to the environment. While a policy of "containing" water on the Ranger mine site was initially adopted, it was envisaged that future procedures may need to include controlled release of waste water. To assist in determining an acceptable operational policy for any such release a research program to collect and assess ecological information (such as description of species, population studies, community metabolism and energy-flow) was initiated. This was necessary as the ecology of the system and of the individual components of the biota were either poorly known or not known at all. It was realized, however, that due to the complexity and variability of this ecosystem it could be many years before such fundamental information could be used to recognize, at an early stage, the onset of adverse effects.

Interest in the Magela Creek and floodplain was also stimulated by its proposed inclusion into Kakadu National Park Stage II, which was eventually gazetted in February 1984. It was also anticipated that, like Kakadu National Park Stage I, gazetted in April 1979, it would be included on the World Heritage List, as eventually occurred.

This paper describes some of the ecological documentation that has occurred since 1978 when investigations were initiated by the Office of the Supervising Scientist for the Alligator Rivers Region and by Pancontinental Limited, the holder of the Jabiluka mining lease adjacent to the floodplain. Aspects considered are water quality and stratification of permanent waterholes, algal diversity and productivity, vegetation distribution and seasonality, and diversity of the native fauna, including aquatic invertebrates, fish, amphibians, reptiles, and birds. The distribution and seasonality of the aquatic vegetation has been discussed in detail to illustrate the seasonal differences that occur.

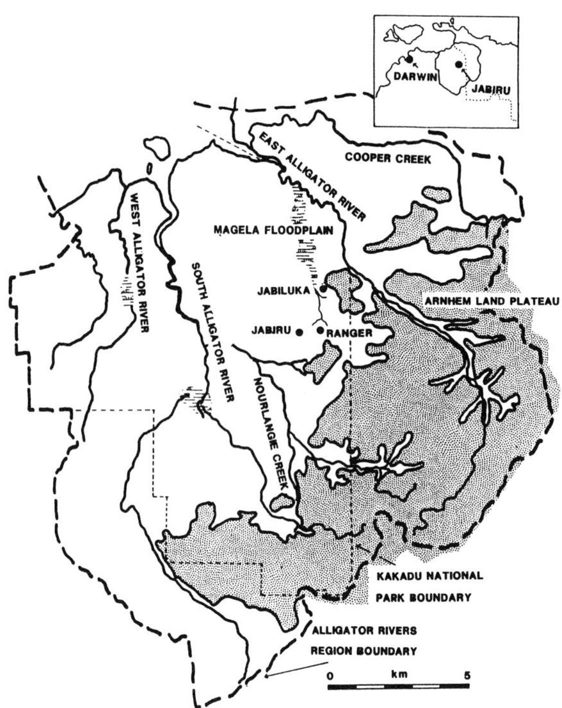

Figure 1. Location of the Magela Creek floodplain and the main features of the Alligator Rivers Region.

The influence of alien species such as the exotic weed Salvinia molesta and the feral introduced Asian water buffalo Bubalus bubalis has not been expanded on, partly due to a lack of localized or information specific to the Magela floodplain.

PHYSICAL CHARACTERISTICS

Location

Magela Creek is situated about 250 km east of Darwin in the Alligator Rivers Region on the western edge of Arnhem Land (Fig. 1). The physical features of the region have been described by Galloway (1976) and Christian and Aldrick (1977). In the south and east is the Arnhem Land Plateau, a deeply dissected, rugged sandstone terrain some 200-300 m above sea level. The edges of the plateau form the Arnhem Land Escarpment which rises up to 250 m above the undulating plains of the lowlands. Extensive floodplains, like that of the Magela, occur along the major rivers and eventually merge into the estuaries.

Creek Hydrology

Magela Creek is a seasonally-flowing tributary of the East Alligator River with a

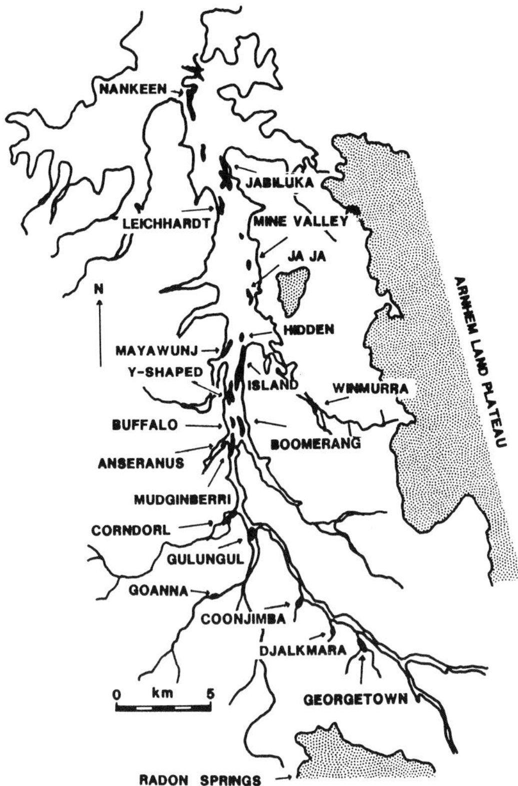

Figure 2. Magela Creek floodplain showing the main billabongs. Fishless and Bowerbird Billabongs are located further upstream and are not shown.

catchment of wooded lowland and sandstone plateau. It consists of five distinct sections: escarpment channels flowing through deep narrow gorges, braided sandbed channels with sandy levees, a series of billabongs (pools in or near the creek rather than oxbow lakes as defined by Bayly and Williams (1973) and connecting channels (the Mudginberri Corridor), a seasonally inundated black-clay floodplain with permanent billabongs and a single channel as it discharges into the East Alligator River. The locations of the billabongs are shown in Fig. 2. Whilst this paper is concerned with all five areas the emphasis has been placed on the black-clay floodplain, referred to simply as the Magela floodplain, and the Mudginberri Corridor.

The braided channels are lined with trees, vary between two and five in number and have a typical width of 200-300 m. Immediately downstream of these channels the Mudginberri Corridor, which is in effect a constriction between opposing lowland slopes, is intermediate in character between the well

Table 1. Billabongs of the Magela Creek system classified by the scheme described by Walker et al. (1984).

Channel Billabongs	Mudginberri Buffalo Boomerang Anseranus Winmurra Island Y-shaped Mayawunj
Backflow Billabongs	Georgetown Coonjimba Gulungul Corndorl Djalkmarra Goanna
Floodplain Billabongs	Hidden Ja Ja MineValley Jabiluka Leichhardt Nankeen

defined upland channels and the downstream floodplain. The areal extent of this floodplain is about 250 km^2. The flood waters recede and evaporate during the dry months, leaving several prominent swamps on the western side and a number of billabongs. Walker et al. (1984)

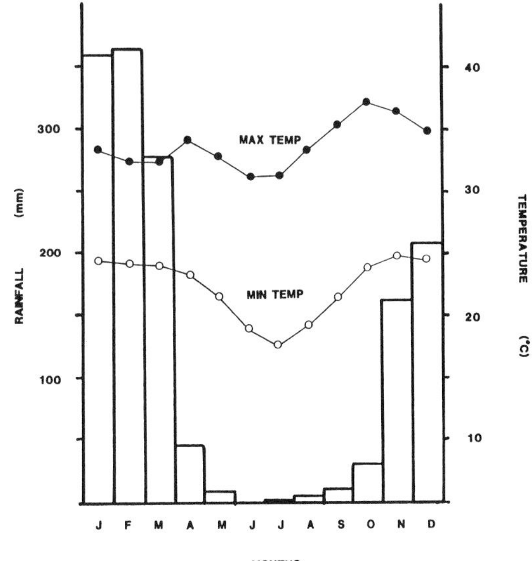

Figure 3. Average monthly rainfall and maximum and minimum temperatures for Jabiru, 1972-84.

have classified the billabongs (Table 1) as channel (depressions in flow channels), backflow (located on small feeder streams and initially filled by water from the main creek) and floodplain (generally remnants of deep channels on the floodplains). The backflow billabongs, in particular, may dry out during the dry season.

Climate

The climate of the region is monsoonal with two dominant seasons, known locally and hereafter referred to as the wet and the dry. The wet generally commences late in the year (November-December) and lasts for 3-4 months; both the onset and duration vary from year to year. The most significant feature of this season is the rainfall from thunderstorms, tropical cyclones and rain depressions. The early part of the wet season is termed 'the build-up' and is characterized by thunderstorms with localized but very heavy rain, whereas thunderstorms during March-April do not produce prolonged rain and indicate the approach of the dry season. The dry season is characterized by south-east trade winds. There is also a high probability of cyclones occurring in March and April.

The average annual rainfall is shown in Fig. 3. The wettest months are January-March, although November and December have over 150 mm on average. When considering a monosoonal climate, however, it is also useful to be aware of daily or individual events. As an example, during November 1984 a single daily recording accounted for 125 mm of the total 224 mm for the month.

The annual range of average maximum temperatures is 31-37°C. The warmest month is October (i.e. in the build-up to the wet) and the coolest period is June-July. The average minimum temperatures are lowest in July (18°C) and highest from November-March (24°C).

A diagramatic representation of the seasonal rainfall and five components of the water level cycle on the floodplain is shown in Fig. 4 (Sanderson et al. 1983). In brief, there is a period of intermittent heavy storms that saturate the soil, followed by more consistent rains which produce creek flow and wide-spread flooding. As rainfall increases continuous flow occurs in the braided channels and eventually into the Mudginberri Corridor before spilling out onto the floodplain. At the cessation of the wet season the creeks stop flowing and the water starts to draw down, leading to cessation of flow and drying of the

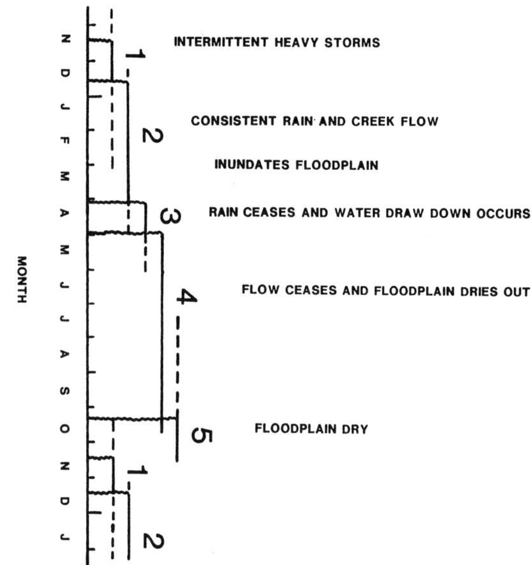

Figure 4. Seasonal cycle on the Magela Creek Floodplain based on observed events from 1979-85 (adapted from Sanderson et al. 1983).

floodplain with isolated pockets of permanent water in billabongs and swamps. The drying-out phase can take many months (phase 4 in Fig. 4). The creeks and channels also dry out completely, or are reduced to a series of water holes or billabongs.

PHYSICO-CHEMICAL WATER CHARACTERISTICS

Water Quality

Magela Creek water, after the first flush of the wet season, is neutral to acidic, very soft with low buffering capacity, and has low ionic concentrations (Brown et al. 1985; Morley et al. 1985; Walker and Tyler 1984). The velocity of the wet season flood front influences the limnological characteristics of the billabongs. In general, a slow advance enables cooler, more acidic and more oxygenated water to displace the warmer, less acidic and less oxygenated water in the billabongs, whereas a higher velocity sheet-type flow initially traps the existing water. The annual degree of change in water quality is also influenced by the duration of the dry season. Solutes generally increase in concentration during the dry season, due to evaporation and ingress of groundwater (Brown et al. 1985).

The initial inflow of acidic water into the billabongs can result in extensive fish kills, such as that reported by Brown et al. (1983) in Ja Ja Billabong. Following a heavy localized storm and

Table 2. Classification of Magela Creek billabongs based on late dry season characteristics (adopted from Walker and Tyler 1984 and Bishop and Forbes 1990).

Type of Billabong	Ionic Character	General Characteristics	Total Phosphorus
Channel	Na/Mg HCO_3	Maintain ionic character of wet season. $K_{25} < 90$ µS cm^{-1} $5.8 < pH < 6.5$ Turbidity <50 NTU	<70 µg l^{-1}
Backflow	NaCl	Progression through dry season to NaCl dominance, $35 < K_{25} < 630$ µS cm^{-1} $4.0 < pH < 6.0$ Turbidity in late dry season > 100 NTU	<400 µg l^{-1}
Floodplain	SO_4	Strong selective concen. of SO_4 plus trend to NaCl. $60 < K_{25} < 1200$ µS cm^{-1}, $3.5 < pH < 5.5$ Turbidity in late dry season > 100 NTU	100-300 µg l^{-1}

significant runoff to the billabong in excess of 3,400 dead fish, mainly Lates calcarifer, Arius leptaspis, Liza diadema and Tandanus ater, were collected. The full extent of the kill was not determined as some fish were removed by predators (birds, crocodiles and monitor lizards). The fish mortality was attributed to the combined effects of low pH values and an increase in aluminum concentration due to natural phenomena, though direct evidence for this has not been obtained.

Billabongs on the Magela were extensively sampled from 1978 to 1981 by Walker and Tyler (1982). A summary of their data is presented in Table 2. During the wet season the billabong water chemistry resembled that of the creek with conductivity less than 20 µS cm^{1}, pH 6.4 - 7.0 and ionic dominances of Na > Mg > Ca > K and HCO_3 > Cl > SO_4. During the dry season the water quality was more site specific with conductivities in mid-July (1980) ranging from 17 - 110 µS cm^{-1} and pH from 5.8 - 7.1. The channel billabongs had conductivities of 17 - 37 µS cm^{-1} and equimolar proportions of Na and Mg, whereas backflow and floodplain billabongs had 38 - 110 µS cm^{-1} conductivities and Na > Mg. Floodplain billabongs, unlike the backflow type, developed Cl anionic dominance. The chemical composition changed later in the dry with conductivities ranging from 18 - 630 µS cm^{-1} and most billabongs becoming Cl dominated. The ionic dominances have been used to classify the billabongs (Table 2).

Mudginberri and Bowerbird Billabongs had equimolar amounts of Na and Mg and low conductivities of 15 - 30 µS cm^{-1} throughout the wet and dry seasons, and were chemically stable after the first flush of the wet season. In contrast Georgetown, Coonjimba, Leichhardt and Buffalo Billabongs progressed towards seawater Na:Cl ratios during the dry season and developed a Na and Cl ionic dominance. Wet season inflows reduced conductivities and restored the Na/HCO_3 ionic proportions. Corndorl, Ja Ja, Mine Valley, Jabiluka, Nankeen and Island Billabongs also tended towards a Na:Cl dominance but, in addition, had a large increase in SO_4, attributed to groundwater ingress. The wet season floodwaters restored the Na and HCO_3 dominance and reduced conductivity.

Nitrogen and phosphorus concentrations during the wet season were relatively low and similar between billabongs, with total phosphorus generally <40 µg l^{-1}, inorganic nitrogen <35 µg l^{-1} and total nitrogen < 1050 µg l^{-1}. Except for channel billabongs these values increased during the dry season (Table 2). The channel billabongs had relatively low nutrient concentrations that changed little with season. Floodplain billabongs were generally hyper-eutrophic by the late dry. The wet season floodwaters flushed and diluted these billabongs. Backflow billabong nitrogen and phosphorus concentrations remained more or less constant through the early dry and increased abruptly late in the dry with total phosphorus eventually exceeding that of the other billabongs.

On the basis of the Vollenweider (1968) nutrient-concentration scheme for classifying the trophic status of waterbodies, the billabongs were regarded, on consideration of the total phosphorus levels, by Walker and Tyler (1982) as meso-eutrophic to hyper-eutrophic during the wet, and all except the channel billabongs hyper-eutrophic during the dry. McBride (pers. comm.), however, has pointed out that total phosphorus values recorded in Ja Ja Billabong by the Northern Territory Water Division over the same period did not exceed 45 $\mu g\ l^{-1}$, designating it as oligo-mesotrophic. On the basis of the inorganic nitrogen concentrations Walker and Tyler (1982) classified the billabongs as ultra-oligotrophic during the wet, while during the dry all except the channel billabongs were mesotrophic to hypertrophic. After considering these data, they suggested that nitrogen and not phosphorus was the likely limiting nutrient. That assessment was later reversed when the phytoplankton levels were correlated to the nutrient levels in the water (Walker and Tyler 1984).

The differences in the trophic status of the billabongs shown by the nitrogen and phosphorus data suggest that the Vollenweider classification may not be appropriate in this situation; a similar suggestion was made for artificial lakes in Queensland by Finlayson and Gillies (1982). The point raised by McBride also suggests the need for caution when assessing the trophic status of the billabongs on total phosphorus concentrations.

Temperature and Dissolved Oxygen

Water temperatures at 10 cm depth in the billabongs vary diurnally with the largest range recorded by Walker et al. (1984) being 9°C in Coonjimba during October. Maximum surface water temperatures were usually around 30°C in open water and about 37°C in thick stands of aquatic plants. The general stratification pattern was classical polymixis with episodic stratification and anoxia in floodplain billabongs like Leichhardt, and ephemeral thermal gradients in the shallow backflow billabongs (Walker and Tyler 1984).

Dissolved oxygen levels were reported by Walker et al. (1984) to have diurnal ranges in excess of 40% ASV (air saturation value) during the dry in productive billabongs such as Jabiluka, Leichhardt and Gulungul. Less productive billabongs such as Bowerbird, had a diurnal range of 20%. At dawn most billabongs had a dissolved oxygen level of about 50% ASV, which corresponded to 3.5 to 5 mg l^{-1}. Diurnal differences were smaller during the wet season. Decomposition of aquatic plants early in the dry causes a heavy oxygen demand and results in anoxic conditions in many billabongs. Noller and Hunt (1985) did not record a diurnal change in flowing creek water with about 6 mg l^{-1} being present at 27°C (76% ASV).

BIOLOGICAL CHARACTERISTICS

Vegetation

Algae

Extensive taxonomic studies of the algae have been carried out with the Bacillariophyceae (diatoms) being investigated by Brady (1979), McBride (1983) and Thomas (1983), and other groups by Ling and Tyler (1986). A preliminary investigation of the Cyanobacteria (blue-green algae), was done by Broady (1984).

The most comprehensive account of the diatoms is the listing of more than 160 taxa from 32 genera of both tropical and cosmopolitan distribution by Thomas (1983). The 160 taxa were divided into four groups depending on whether they were confined mainly to the escarpment, the lowlands and the floodplain, or were taxa with no distinguishable distributional pattern. The diatom flora of the escarpment pools varied less than that of the floodplain billabongs, but more sampling would be required to determine how wide spread many of the taxa are. Overall, the diatom flora was considered very rich, allied to the flora of south-east Asia, and spatially and temporally diverse. Annual changes in water quality and the variable nature of the environment contribute to this diveristy. The backflow billabongs were particularly rich in taxa.

Ling and Tyler (1986) recorded over 530 taxa from groups other than the diatoms, of which 66 were new to science or could not be identified. About 360 taxa not previously recorded from the area were collected. The Desmidiaceae (desmids) were particularly rich floristically; a situation common in the tropics. Like the diatoms, the species recorded in this study were similar to those of south-east Asia.

The Mudginberri Corridor, however, was regarded by McBride (1983) as having a low diatom species diversity. Species occurrence and diversity did, however, vary from the wet to the dry season, and spatially in the upper half meter of water due to diurnal temperature fluctuations. The biomass of diatoms attached to aquatic macrophytes was equivalent to about 30% of the macrophyte biomass. In the Mudginberri Corridor this amounted to about 16,500 t dry weight of macrophytes and 5,000 t of epiphytic diatoms.

Broady (1984) investigated the widespread desiccated crusts and felts on the floodplain soil during the dry season and recorded Microchaete and Scytonema species not recorded by Ling and Tyler (1986). There is no information on the possible nitrogen fixing role played by the heterocystous genera - Scytonema, Calothrix, Hapalosiphon and Stigonema - in the nitrogen economy of the floodplain. The mucilaginous sheaths secreted by these species combined with the intertwining filaments do, however, act to stabilize the soil surface and prevent moisture loss (Broady 1984).

A characteristic feature of the floodplain during the dry season is the development of large patches of green and red algal scums on the remaining pools of water. The scums consist of phytoflagellates, in particular Pyramimonas, Chlamydomonas, Chlorogonium and Euglena species (Kessell and Tyler 1982). Euglena sanguinea, a large species, is responsible for the commonly observed red color, but changes in species composition can cause a change to green.

Walker and Tyler (1983) investigated the primary productivity of phytoplankton in Magela Creek billabongs. The euphotic zone varied from 0.7 to >6.5 m depth, with the lower values being associated with increased turbidity during the dry season. The higher values were in Bowerbird Billabong, an escarpment rockpool with extremely clear water (Walker et al. 1982). Chlorophyll a values were not high when compared to other tropical lakes. During a Microcystis aeruginosa bloom in Gulungul Billabong a maximum value of 150 mg m^{-3} was recorded, whereas in Bowerbird Billabong the maximum recorded was 5 mg m^{-3}. Higher values were recorded in other billabongs by Kessell and Tyler (1982). Productivity was low during the wet, increased during the dry but was reduced again if turbidity increased, as occurred in floodplain and backflow billabongs. Island and Leichhardt Billabongs were the most productive with dry season values generally above 150 mg O_2 m^{-3} h^{-1}.

Kessell and Tyler (1982) recorded between 80 and 100 taxa in billabongs on all sample occasions. The population composition and number could, however, change in a matter of hours with considerable spatial heterogeneity, both vertically and horizontally. This led to considerable heterogeneity in chlorophyll levels. In Leichhardt Billabong a very broad daily range of 20-1500 mg m^{-3} was recorded. A general pattern of uniform vertical distribution of chlorophyll in the morning, concentration at 0.5-1 m depth during the afternoon and dispersion again in the evening was observed.

Vegetation Distribution

The Magela Creek floodplain contains about 75 known species of water plants from 29 families. In addition there are about 150 trees, shrubs, herbs and vines not covered by the definition of Cook et al. (1974) for water plants, that is, "...Pteridophyta and Spermatophyta, where photosynthetically-active parts are permanently, or, at least for several months each year, submerged in freshwater or floating on the water surface." The main families and number of known water plant species are - Cyperaceae (10), Poaceae (8), Hydrocharitaceae (7), Menyanthaceae (5), Nymphaeaceae (4) and Lentibulariaceae (4).

The earliest description of the Magela Creek vegetation was by Specht (1958) who recognized three broad communities - tall Melaleuca leucadendra (sensu lato) association, Eleocharis dulcis association, and E. dulcis/ Nymphaea gigantea var. violacea association. Story (1976) described and mapped similar broad categories. Williams (1979) in seeking to establish a relationship between vegetation and patterns of water flow, distinguished and described six vegetation types on the floodplain. These were regarded as direct indicators of water depth. However, as the survey times and aerial photographs used did not coincide with times of peak wet season vegetation (April-May) they did not distinguish the boundaries of communities dominated by annual species such as Oryza meridionalis and Hygrochloa aquatica.

Morley (1981) classified and mapped peak wet season (April) herbaceous aquatic vegetation on the floodplain south of Nankeen Billabong. Distribution of the 36 communities recognized was largely attributed to flow conditions and water depth. This classification, however, was not reproducible in subsequent wet seasons (Sanderson et al. 1983). A classification with fewer categories was subsequently adopted.

The community descriptions of Sanderson

et al. (1983), however, did not accommodate year to year changes. Community structure and species dominance have been observed to change markedly from one year to the next. This may be due, in part, to the impact of early-wet season storms and dry periods on the germination and establishment success of the annual species that form 60 - 80% of some plant communities (Taylor and Dunlop 1985). There also appears to be a succession in dominance after flooding and during the drying out phase on the floodplain. In an attempt to present a generalized and broad vegetation classification Finlayson et al. (1989) used peak wet season vegetation data from several years to describe and map the floodplain flora. The presence of species reaching their biomass peak at other times of the year was also considered. Each mapping unit was named according to observed wet season to early-dry season vegetation or given a general descriptive name. The indicator species chosen may not be present throughout the complete annual cycle, but is the main species present at the time of greatest plant biomass and diversity. The resultant map (Fig. 5) is therefore an attempt to take into account the temporal variations of species dominance and did not delineate the minor plant communities identified by Sanderson et al. (1983). The major plant communities (or mapping units) are described below and discussed in terms of the seasonal wet-dry cycle, presented in Fig. 4, that occurs on the floodplain. Changes in associated species are also discussed to provide a broad basis to this vegetation classification.

(i) Melaleuca open forests and woodlands (canopy cover 10-70%) cover 7390 ha and are dominated by one or more Melaleuca species (M. viridiflora and M. cajaputi around the floodplain edges and at the northern end of the floodplain, and M. leucadendra in back swamps inundated for 6 - 8 months). The understory varies and includes Nymphaea violacea, N. sp. and Pseudoraphis spinescens and/or Hymenachne acutigluma, and/or O. meridionalis, Salvinia molesta and Najas tenuifolia in the wet season (periods 2 and 3 in Fig. 4). During the dry season (periods 1, 4 and 5, Fig. 4) the understory is dominated by P. spinescens and/or Hygrochloa acutigluma or annual terrestrial herbs.

(ii) Melaleuca open woodlands (canopy cover < 10%) cover 1290 ha and are dominated by M. leucadendra in areas inundated for long periods, often with a dense understory of P. spinescens, H. acutigluma or O. meridionalis. E. dulcis may be present, while annual herbs may replace O. meridionalis during the dry.

(iii) Mixed swamp covers 2090 ha of the floodplain. Some swamps dry out regularly while others do so infrequently, a factor that greatly influences the species dominance. Floating mats of the perennial grasses H. acutigluma and Leersia hexandra with Ludwigia

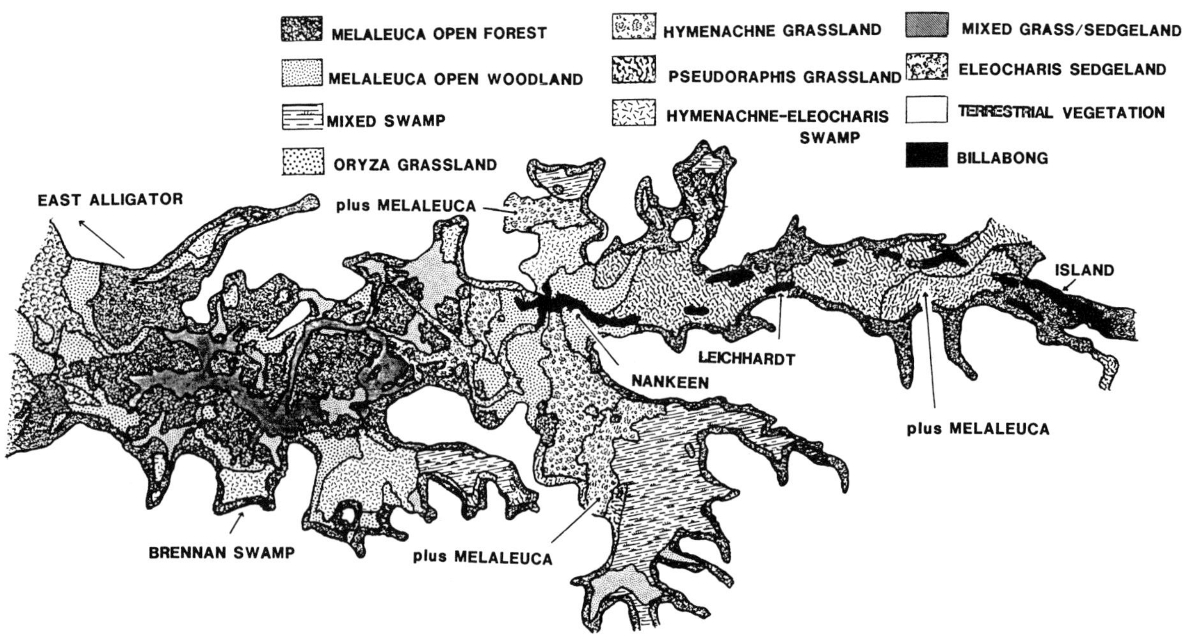

Figure 5. Vegetation map of Magela Creek floodplain (1:100, 000 scale).

adscendens and the mat colonizing sedge Cyperus platystylis are commonly present. The small floating species Azolla pinnata and Spirodela polyrhiza also occur. In permanently flooded swamps the tall sedge Hymenochaeta grossa forms dense clumps during the late dry season. The perennial, emergent or floating leaved lily Nelumbo nucifera can form very dense stands. Nymphoides indica, Nymphaea spp. and Utricularia spp. commonly occur in more open water areas early in the dry season. Chara sp. becomes increasingly abundant at the end of the dry season. Other species present include M. leucadendra, Eleocharis sp. and Vallisneria sp.

(iv) Oryza grassland covers 2730 ha and is dominated by O. meridionalis at the end of the wet season. In the dry season they are largely barren or have senescent O. meridionalis with persistent Phyla nodiflora, L. adscendens in a xerophytic form, and P. spinescens. Early-wet season storms (period 1, Fig. 4) cause germination of O. meridionalis, Digitaria sp., Aeschynomene spp., H. aquatica, Heliotropium indicum and Coldenia procumbens. The pattern of storms is influential on the subsequent success of these species (Taylor and Dunlop 1985). Once flooding (period 2, Fig. 4) occurs aquatic species such as Maidenia rubra, Isoetes muelleri, Blyxa spp., Nymphoides spp., Nymphaea hastifolia and Eleocharis sp. may appear. During the drying out phase (period 4, Fig. 4) L. adscendens, Ipomoea aquatica and Utricularia spp. increase in abundance while herbs eg. Commelina lanceolata are common on the wet mud.

(v) Hymenachne grassland covers 1930 ha and is dominated by H. acutigluma throughout the year. Minor species present include Aeschynomene spp., L. adscendens, S. polyrhiza, A. pinnata, Lemna minor, N. violacea, O. meridionalis and P. spinescens. The naturalized exotic grass Urochloa mutica appears to be invading this community.

(vi) Pseudoraphis grassland is dominated by the perennial, emergent grass P. spinescens which has a turf-like habit during the dry season and grows up through the water during the wet season. Other aquatic species such as Nymphaea violacea, N. macrosperma, N. tenuifolia, H. aquatica and Ultricularia spp. become more abundant at the end of the wet season (period 3, Fig. 4). In some years P. spinescens may fail to assume complete dominance; Nymphaea spp., Eleocharis spp. and Nymphoides indica, often followed by dry season annuals such as Fimbristylis aestivalvis, Cyperus digitatus and Glinus oppositifolius become locally dominant.

(vii) Hymenachne-Eleocharis swamps cover 1290 ha and occur in areas that dry out seasonally and are dominated by H. acutigluma or the slower to establish sedge Eleocharis sp. Long periods of shallow water conditions appear to favor the latter.

(viii) Mixed grasslands and sedgelands cover 1120 ha of the floodplain. A mixture of species - O. meridionalis, P. spinescens and Eleocharis spp. is arranged in a mosaic pattern on the northern end of the floodplain. Variations in localized drainage caused by undulating topography and flow channels appear to influence the pattern.

(ix) Eleocharis sedgelands cover 960 ha and occupy shallow flooded areas near the merging of the Magela floodplain to that of the East Alligator River. Eleocharis sp. dominates during the wet but is replaced by annual herbs such as G. oppositifolius, C. procumbens, Phyla nodiflora, H. indicum and Cardiospermum halicacabium that are dominant during the dry season.

(x) Open-water communities are restricted to 160 ha of the floodplain. Permanent billabongs, flow channels and seasonal, shallow waterholes contain the lily N. macrosperma and a number of submerged species. The latter includes N. tenuifolia during the wet season, changing to Ceratophyllum demersum and Hydrilla verticillata during the dry. The western banks of the billabongs are generally fringed with the trees Barringtonia acutangula, Pandanus aquaticus, Melaleuca spp. and the vines Aniseia martinicensis and Merremia gemella. Associated with this fringe is a floating mat of L. hexandra, H. acutigluma and L. adscendens and the alien species U. mutica and S. molesta. S. molesta may be dominant on billabongs during the dry.

Seasonal Changes in Vegetation

An outstanding feature of the floodplain flora is the variation in floristic composition and foliage cover between the wet and dry seasons (Finlayson et al. 1989). The abundance and diversity of the wet season vegetation contrasts with the sparse cover of the dry season herbs, grasses and sedges. Of the 222 plant species recorded from the Magela system five are restricted to permanent billabongs and swamps. The remaining 217 species includes 94 which occur in seasonally inundated areas and 158 that occur in the Melaleuca forests that fringe the open plain area. Many of the latter are terrestrial species that do not survive in a vegetative form during the wet season. The success of the majority of species depends on

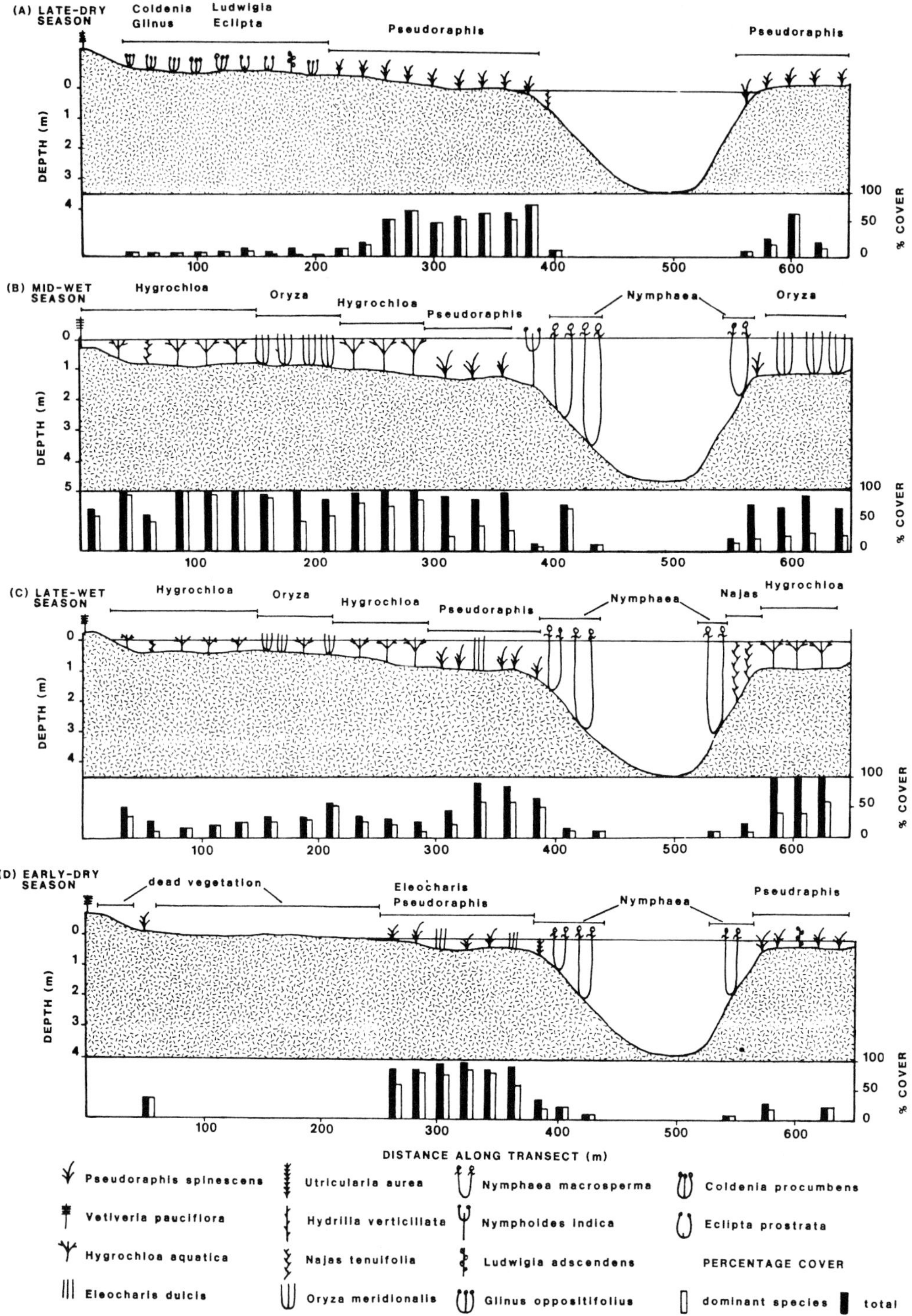

Figure 6. Seasonal changes in floodplain vegetation, percentage cover and dominant biomass, along transects at Nankeen Billabong.

mechanisms that enable them to survive the extremes of a tropical wet-dry climate; dry season drought and wet season inundation.

The floodplain plants have been divided into 3 groups on the basis of life histories:

(i) perennials;
(ii) geophytic perennials;
(iii) annuals;

Sixty-eight perennial species occur on the floodplain, including the emergent aquatic grasses P. spinescens and H. acutigluma, and the paperbark trees Melaleuca spp. Five Melaleuca species occur on the floodplain with a major division between species occurring in seasonally inundated areas (eg. M. viridiflora) and those occurring in areas subjected to prolonged waterlogging (eg. M. leucadendra). The latter develop aerial adventitious roots which can form substantial trunk buttresses.

Thirty-four of the perennials are terrestrial, 26 are aquatic and 8 others are difficult to classify. The aquatic species are dominated by herbs (12 spp.) and grasses (5 spp.).

Geophytic perennials are species with underground perennating organs and are generally confined to seasonally inundated areas. The 14 species in this group include four Eleocharis and four Nymphaea species.

One hundred and thirty-nine species are considered as annuals; 37 of these are water plants and 102 are terrestrial herbs, grasses and sedges. The terrestrial species, while being relatively diverse, do not constitute a major portion of the total annual standing crop on the floodplain. In seasonally inundated areas they depend on a carryover of seed in the soil to survive the dry season drought.

The scheme prepared by Sanderson et al. (1983) after 4 years of observation, reproduced in Fig. 4, depicts 5 hydrological periods important to the seasonal growth of plants on the floodplain. The results of surveys across Nankeen Billabong and the adjacent floodplain (Fig. 6) are used to illustrate the dynamic nature of the vegetation in the context of this scheme.

Most of the floodplain is dry at the end of the dry season (Fig. 6a) except for water in permanent billabongs, channels and swamps. Terrestrial species and water plants that have a "terrestrial" growth form dominate the vegetation on the open plain. Intermittent heavy rain storms saturate the soil, increase the vigour of the terrestrial species, and allow some water plants (eg. H. aquatica) to germinate. Consistent rains cause Magela Creek and other tributaries to flow and inundate the floodplain. The flood conditions eventually cause the terrestrial species to die and water plants to germinate and grow. By mid to late-wet season (Fig. 6b) aquatic plant communities are well established.

P. spinescens grows rapidly as water levels increase, attaining a peak standing crop of about 1.4 kg m^{-2} (Finlayson 1988). The annual floating-leaved grass H. aquatica or the submerged rooted N. tenuifolia can locally displace P. spinescens. These species, however, are generally less productive with standing crops less than 0.5 kg m^{-2} (Sanderson et al. 1983). Towards the end of the period of inundation the water lilies N. violacea and N. macrosperma become widespread. N. violacea does not occur in high densities though it is a conspicuous species in seasonally inundated areas and single plants can weight 0.5 kg dry weight (Sanderson et al.1983). N. macrosperma is confined to permanent waterbodies and forms dense fringes around some billabongs; as many as 66 plants per 50 m^2 have been recorded by Sanderson et al. (1983).

At the end of the wet season creek flow is greatly reduced and the water level drops rapidly. Once all flow has ceased drying out is due entirely to evaporation. The aquatic plants gradually senesce and decline, as indicated by the lower cover values in Fig. 6c. A large amount of debris is deposited among species like P. spinescens (Fig. 6d,a) that are able to establish a "terrestrial" growth form to survive through the dry season.

The profiles in Fig. 6 illustrate the seasonal nature of the floodplain vegetation, but of equal importance are the spatial changes that occur between years. For example, the communities at Nankeen have since 1982 shifted towards an O. meridionalis dominance (Fig. 5); possibly due to the influence of early wet season storms on the germination of annual species. The variability in timing and hydrological nature of the flood regime encountered each year is important in determining the composition of the plant communities and may even cause large changes to community distributions. The full extent of this influence is not well understood in relation to the germination and growth characteristics of the plant species.

Fauna

Aquatic Invertebrates

The first scientific collection of macroinvertebrates from Magela Creek billabongs was by Marchant (1982a). About 90 taxa were identified over one year from 5 billabongs. In the backflow billabongs (Coonjimba, Georgetown and Goanna) the greatest number of taxa (about 45) and individuals collected (about 1000 per minute) occurred during the late wet and early dry season period. Channel billabongs had less marked seasonal fluctuations. The presence of high numbers of taxa and individuals at the end of the wet season was attributed to the food and shelter available from the well developed aquatic plant communities at that time.

During the dry season survival strategies vary, with animals of some taxa hibernating in the mud (Gastropoda), having resistant eggs (Ephemeroptera) or pupae, as certain insect species do. Adults of some true-aquatic species may also survive in low densities in the deeper channel billabongs, but not in the shallower backflow billabongs. At the beginning of the wet season there is a rapid resurgence of macroinvertebrates. Some insects, like Cloeon fluviatile and Tasmanocoenis sp. have complete life cycles of less than one month (Marchant 1982b).

The microcrustacea of the billabongs were regarded by Julli (1986) as being a rich assemblage with 43 Cladocerans and 9 Copepod species. Of these, 37 Cladoceran and 5 Copepod species were collected from among the aquatic vegetation, the more common being Ephemeroporus barroisi, Echinisca triserialis, Ilyocryptus spinifer, Mesocyclops sp. nov. and Microcyclops varicans. As the vegetation declined throughout the dry, area of littoral habitats decreased and these taxa became less common. Open water assemblages are able to reform with common species being the calanoid Copepods Diaptomus lumholtzi and Calomoecia ultima, the cyclopoid Copepods Mesocyclops decipiens and M. leuckarti, and the Cladocerans Diaphanosoma excisum, Ceriodaphnia cornuta, Moina micrura and Bosminopsis deitersi.

The cyclopoid Copepods were normally the first to appear after the billabongs were flushed by wet season floods. They were followed by the calanoid Copepods and then, as turbidity increased during the dry season, by the limnetic Cladocerans.

Tait et al. (1984) carried out a two year study of the plankton and littoral microfauna of 8 Magela Creek billabongs and presented a species checklist. The list (excluding the Protozoa) contained 227 Rotifers, 14 Copepods, 35 Cladocerans and 5 Ostracods. A large proportion of the species were littoral or epiphytic, and there were vast spatial and temporal differences in species abundance. Species diversity increased during the wet season and decreased during the dry season.

Fishes

A study of the fishes of the Alligator Rivers Region was done by Bishop et al. (1981). They described the main habitats occupied by fish and reported on the abundance and fresh weight of the main species. Abundance and fresh weight were determined by standardized observations and gill netting techniques. All abundance and weight references in the following discussion refer to these techniques. Bishop and Forbes (1986) have reported more generally on the freshwater fishes of northern Australia, including information on the Magela Creek system. The information presented below is taken from these two reports. The habitats of the Magela Creek system are divided into plateau, escarpment and lowland categories. A species list is given in Table 3.

The plateau habitat consists of spectacular cascades and swirling pot-hole pools that become isolated during the dry season. The most abundant fish there were Melanotaenia nigrans and Leipotherapon unicolor.

The rock pools in the upper reaches of the creek in the escarpment are large, turbulent, connected pools during the wet season, but become isolated by steeply shelving bedrock and vertical drops during the dry. Pools in close proximity to the lowlands have more diverse fish communities than those immediately below large waterfalls in the upper reaches of the creek. The most numerous species in the lower reach pools were Craterocephalus marjoriae, Craterocephalus stercusmuscarum, Ambassis macleayi, Melanotaenia splendida inornata and Amniataba percoides. Seasonal changes in abundance generally followed changes in the number of C. marjoriae with more present in the early-wet season. The major proportion of the weight of fish caught in Bowerbird Billabong was due to Tandanus ater, A. percoides, Nematalosa erebi, Syncomistes butleri and Scleropages jardini with a decrease from 7 kg fresh weight late-dry to 1.5 kg in the wet season.

The escarpment also contains spring-fed perennial streams such as that at Radon Springs which contained 10 fish species dominated by M. nigrans, Neosiluris hyrtlii, Pingalla midgleyi

Table 3. Fish species of the Magela Creek system (from Bishop et al. 1981 and Walden pers. comm.).

Megalops cyprinoides	tarpon
Nematalosa erebi	bony bream
Scleropages jardini	saratoga
Hexanematichthys leptaspis	fork-tailed catfish
Tandanus ater	butter jew
Neosilurus hyrtlii	Hyrtl's tandan
Copidoglanis rendahli	Rendahl's tandan
Strongylura kreffti	long tom
Melanotaenia nigrans	black-striped rainbow fish
Melanotaenia splendida inornata	checkered rainbow fish
Craterocephalus marjoriae	fly-specked hardyhead
Craterocephalus stercusmuscarum	Marjorie's hardyhead
Pseudomugil tenellus	blue-eye
Pseudomugil gertrudae	spotted blue-eye
Ophisternon mitturale	one gilled eel
Ambassis macleayi	reticulated perchlet
Ambassis agrammus	sailfin perchlet
Denariusa bandata	penny fish
Lates calcarifer	barramundi
Amniataba percoides	banded grunter
Hephaestus fuliginosus	black bream
Leiopotherapon unicolor	spangled grunter
Syncomistes butleri	sharp nosed grunter
Pingalla midgleyi	black anal fin grunter
Glossamia aprion	mouth almighty
Toxotes chatareus	archer fish
Liza diadema	mullet
Glossogobius giurus	flat-headed goby
Hypseleotris compressus	carp gudgeon
Mogurnda mogurnda	purple spotted gudgeon
Oxyeleotris lineolatus	sleepy cod

and L. unicolor. In general, they were more common in the mid-wet.

The sandy creekbeds of the lowlands contain about 21 fish species with fewer at sites that dry out completely during the dry season. The most abundant species are C. marjoriae, M. maculata, Ambassis spp. and C. stercusmuscarum. Fish density increases in the late-wet to early-dry period and then decreases. Fresh weight of fish caught was generally no more than 10 kg with T. ater, Arius leptaspis, N. erebi, Toxotes chatareus, Strongylura kreffti and L. unicolor all contributing.

The muddy backflow or billabongs of the lowlands can have high flow conditions during the wet season and may even become part of the large inundated backflow section of the system. The number of fish species varies from 13 in Fishless Billabong to 22 in Djalkmara and Coonjimba Billabongs. The most abundant generally include Ambassis agrammus, A. macleayii, M. maculata, Denariusa bandata, Glossamia aprion, A. rendahli and N. erebi. The two Ambassis species, M. maculata and G. aprion contribute most to fish abundance throughout the year, with a peak in the wet season. The fresh weight of fish caught though, was dominated by Arius leptaspis, N. erebi, T. ater, Oxyeleotris lineolatus and P. rendahli, with some variation between billabongs. Most of this catch, from 2-17 kg in the different billabongs, was caught either during the late-dry or in the wet season.

Billabongs in the Mudginberri Corridor contained 23-25 fish species with the most abundant being Ambassis agrammus, C. stercusmuscarum, M. splendida inornata, Glassogobius giuriis, D. bandata, A. macleayi and G. aprion. Peak abundance, usually A. macleayi or G. aprion, occurred either in the late-wet or early-dry. Total weight of fish caught was much higher than in other habitats, reaching 31 kg in Island Billabong, due mainly to Arius leptaspis and T. ater. This decreased during the middle of the wet season, probably due to large fish leaving the billabongs.

The upper floodplain billabongs such as Ja Ja, Leichhardt and Nankeen do not receive water flow as early in the wet season as habitats upstream. They contained 22-24 fish species dominated in abundance by Ambassis agrammus, C. stercusmuscarum, M. splendida inornata, Hexanematichthys leptaspis, D. bandata and A. macleayi. The amounts caught in the gill nets were high, 19-29 kg, due mainly to Arius leptaspis and T. ater, compared to 3 kg minimum values.

The information on fish communities of different habitats suggests a relationship between habitat and fish species. Smaller juveniles are most diverse and abundant in the nursery habitats - the muddy lowland and floodplain lagoons. Larger juveniles and smaller adults are most abundant and diverse in the corridor billabongs and sandy creekbeds that connect the permanent floodplain and the muddy lowland billabongs during the wet season. Larger adults are most diverse in floodplain and muddy lowland billabongs and to a lesser extent in the escarpment main channel.

The most prominent effect of fish movement throughout the system is the

Table 4. Frogs of the Magela Creek system in the main natural habitats (after Tyler et al. 1983).

Species	Natural Habitat
Litoria bicolor	billabong fringes, flooded grasslands
Litoria caerulea	escarpment streams and pools
Litoria coplandi	escarpment streams and pools
Litoria dahlii	floodplain, billabong fringes
Litoria inermis	billabong fringes
Litoria sp. near latopalmata	billabong fringes, flooded grasslands
Litoria meiriana	escarpment streams and pools
Litoria microbelos	floodplain, billabong fringes
Litoria nasuta	grassland fringes, sclerophyll forests
Litoria personata	escarpment streams and pools
Litoria rothi	billabong fringes, flooded grasslands
Litoria rubella	billabong fringes, flooded grasslands
Litoria tornieri	billabong fringes, flooded grasslands
Litoria wotjulumensis	escarpment streams and pools
Cyclorana australis	billabong fringes, flooded grasslands
Cyclorana longipes	billabong fringes, flooded grasslands
Limnodynastes convexiusculus	billabong fringes, flooded grasslands
Limnodynastes ornatus	escarpment streams and pools, grasslands
Megistolotis lignarius	escarpment streams and pools
Notaden melanoscaphus	grassland, sclerophyll forest
Ranidella bilingua	billabong fringes, flooded grassland
Uperoleia arenicola	escarpment streams and pools
Uperoleia inundata	grassland, sclerophyll forest
Spherophryne robusta	escarpment, billabong fringes

recolonization of sandy creekbeds and muddy billabongs on the lowlands during the wet season. These are feeding and breeding areas and are entered mainly from downstream. Migratory behavior appears to be a survival strategy of many of the fish species. During the wet season the fish communities become more homogeneous as water flows link the dry season refuge areas and enable more species to disperse and breed. Many species migrate upstream in "waves" that may be species specific. The most notable change from the dry to the wet season is a shift from Ambassis agrammus to M. maculata dominance. Explanations for these changes are lacking, though migration apparently induces considerable community change.

Amphibians

Twenty four species of frogs from 9 genera (Table 4) were identified in the Magela Creek system (Tyler et al. 1983). In general they are totally inactive during the dry season with many species remaining beneath the ground. Substantial rainfall is needed to soften the soil to enable release of some species. At this stage temporary pools for breeding purposes are likely to have formed. Species that do not bury themselves during the dry but simply seek shelter to avoid dehydration are equally dependent on these pools for breeding.

The density and movement of frog species on the floodplain has been described by Tyler and Crook (1987). The greatest density occurs on poorly drained sandy soils with the lowest on well drained gravelly soils. Cyclorana australis is very common and is found in swamp grass/sedgeland and open woodland whereas Litoria nasuta, L. sp. nr. latopalmata, L. inermis, L. tornieri and L. wotjulumensis are common in swamps.

Density estimates point to the common occurrence of species like C. australis, but such values are inherently biased by the sampling techniques and the habits of the species themselves.

During the dry season species such as L. inermis, L. sp. nr. latopalmata, L. tornieri, L. nasuta, C. longipes and R. bilingua were sighted around natural water bodies. During the day they sheltered in cracks in the dry mud or in hollows beneath logs or leaf litter.

Changes in the physical environment affect the spatial distribution and abundance of the frogs. Heavy rains bring species like L. dahlii from their hibernation sites to the surface. As heavy rain continues they appear to disperse overland to flooded wallows, buffalo pug marks and shallow depressions that offer spawning sites. Banks around billabongs like Nankeen act as refuges from flood waters for adult R. bilingua and L. nasuta and juvenile L. nasuta and L. sp. nr. latopalmata. As water levels

Table 5. Large reptiles of the Magela Creek system (from various sources).

Varanus gouldii	sand goanna
Varanus panoptes	sand goanna
Varanus mertensis	water goanna
Varanus mitchelli	water goanna
Acrochordus arafurae	filesnake
Crocodylus porosus	estuarine crocodile
Crocodylus johnstoni	freshwater crocodile
Bothrochilus fuscus	water python
Amphiesma mairii	freshwater snake
Stegonotus cucullatus	slatey-grey snake
Enhydris polylepis	Macleay's water snake
Pseudechis australis	king brown snake
Chelodina rugosa	northern long-necked turtle
Elseya dentata	northern snapping turtle
Carettochelys insculpta	pitted-shell turtle
Emydura australis	boof-head turtle
Emydura victoriae	northern short-necked turtle
Elseya latisternum	saw-shelled turtle

rise these sites are inundated and the frogs only survive by sheltering in flooded vegetation. The arboreal species L. rothi, L. bicolor and L. rubella shelter in Pandanus aquaticus and Barringtonia acutangula.

Reptiles

The reptiles of the upper Magela Creek area have been investigated by Sadlier (1981). In general the upper sections of the creek were not suitable habitats, except for arboreal species, while along the mid-reaches more ground dwelling species were found. Access to the surrounding woodland is important to enable escape from inundated areas during the wet season. A list of the major aquatic or floodplain species is presented in Table 5.

The food habits, habitats and general biology of five species of reptiles - filesnakes (Acrochordus arafurae), sand goannas (Varanus panoptes and V. gouldii) and water goannas (V. mertensi and V. mitchelli) - on the Magela system were investigated by Shine (1986).

The file snake is only distantly related to other living snakes. It is large and heavy bodied with small scales covering the entire body and a loose, baggy skin. Females grow larger than males and usually have heavier bodies, shorter tails and larger heads.

During the dry season file snakes are mainly restricted to the permanent floodplain billabongs, congregating around the fringing tree Pandanus aquaticus and floating grass mats (Shine and Lambeck 1985). There are more females than males with more than 1,000 adult females in each of the floodplain billabongs. The channel billabongs do not contain as many snakes. In the wet season they move out of the billabongs to shallow inundated grasslands and among roots of the freshwater 'mangrove' Barringtonia acutangula.

The sand goannas feed mainly on reptiles and mammals. Lizards and crocodile eggs are important food for V. gouldii. Mammalian prey include the small species of Mus, Rattus, Pseudomys, Sminthopsis and Isodon with larger macropods being taken as carrion. Insects are also consumed. Varanus mertensis has a less diverse diet with a high proportion coming from freshwater crabs. Varanus mitchelli which can forage on land and underwater, consumes more fish and orthopterans. The large sand goanna V. panoptes favors the riparian habitats but can be found almost anywhere on the Magela system.

The two Australian species of crocodiles, the small (up to 3 m in length) freshwater Crocodylus johnstoni and the larger (up to 7 m in length) "estuarine" Crocodylus porosus are found throughout the Magela system. C. johnstoni is endemic to Australia and is restricted to freshwater billabongs, floodplains and rivers. It nests in the dry season with 10-20 eggs laid in a cavity excavated in sand, has a fish and insect diet and is relatively poorly studied (Webb 1977). The larger C. porosus is found in both tidal and freshwater sections of rivers. It nests just above water during the wet season on the edge of billabongs, in freshwater swamps or on river sides; flooding can be catastrophic to a successful nesting season (Webb 1977).

Smaller C. porosus are opportunistic feeders and eat mainly invertebrates, whereas the larger animals eat more vertebrates including birds, wallabies and other crocodiles as well as other reptiles (Taylor 1979). The diet and condition of these animals varies between habitats.

The first census of C. porosus in the Magela was by Messel et al. (1979). While 26 individuals were detected in the billabongs very few nests were found. Grigg and Taylor (1980) in a later survey similarly found few good nesting areas on the floodplain where much of the vegetation in suitable areas had been severely damaged by trampling and grazing of feral buffalo. Coupled with the lack of permanent water the floodplains were considered poor nesting sites in comparison to the permanent billabongs where some nests were found. During the wet season crocodiles move from rivers such as the East Alligator to the floodplains, with some remaining in favored localities, such as billabongs and permanent swamps, after the wet season (Jenkins and Forbes 1985).

Populations of turtles in the Alligator River Region were sampled by Legler (1980), primarily for dietary analysis. Four species were dissected and analyzed for stomach contents. Chelodina rugosa is carnivorous, consuming small fish and crustaceans. Elseya latisternum is omnivorous while Elseya dentata and Emydura australis are herbivorous. Elseya dentata generally consumed plant material of terrestrial origin that had fallen into the water, whereas E. australis consumed aquatic plant fragments. A fifth species, Carettochelys insculpta is relatively unknown in Australia, though its occurrence has been substantiated in the South Alligator River drainage area by Legler (1980).

Birds

A general account of birds of the monsoonal area of the Northern Territory has been prepared by Morton and Brennan (1990). The floodplains of the Alligator Rivers Region are important habitats for waterbirds, providing feeding sites during the dry season and breeding grounds during the wet season. About 35 species of migratory birds visit this region from the northern hemisphere and southern Australia. The most common are Numenius minutus, Charadrius veredus, C. leschenaultii, and C. mongolus. They tend to reach their peak population both immediately prior to and just after the wettest months (January-March) of the wet season. As the floodplain waters recede a number of waterfowl species from southern Australia, Anas gibberfrons, Malacorhyncus membranaceus, Athyra australis and Porphyris porphyrio are common.

The water bird fauna is diverse with two main groups - the aquatic herbivorous anatids and the piscivorous/insectivorous ardeids. The former include Anseranus semipalmata, Dendrocygna arcuata and Nettapus pulchellus while the latter include Egretta garzetta, E. intermedia, E. alba, and Ardea picata. The diets of all species are linked to the phenological state of the floodplain plants. N. pulchellus feed on H. aquatica during the mid-wet and change to Limnophila indica, Caldesia oligococca and Nymphoides spp. In the early dry N. violacea becomes more important while N. macrosperma and Hydrilla verticillata are important when the habitat has contracted further in the late-dry. The piscivorous/insectivorous species have different habitat and foraging strategies that affects the types of prey likely to be encountered.

Waterbird use of the floodplains varies seasonally (Morton et al. 1984) with the Magela having a peak population in the late dry when water and food resources are scarce elsewhere. Approximately 200,000 waterbirds occupy the floodplain at this time, decreasing to about 50,000 during the wet season. A list of common waterbirds of the Magela floodplain is given in Table 6.

ENVIRONMENTAL CONSIDERATIONS

In this study part of the documentation of the Magela Creek system is reviewed; particular attention is focussed on the black-soil floodplain section. The stimulus for such documentation resulted from the Ranger Uranium Environmental Inquiry (Fox et al. 1977) which recognized public concern over potential pollution threats associated with the development of the Ranger uranium mine. A recommendation of the inquiry was that a research institute be established. The Alligator Rivers Region Research Institute was established under the Alligator Rivers Region (Environmental Protection Act 1978), with a mandate 'to conduct, co-ordinate and integrate research required to ensure the protection of the environment from any harmful consequences resulting from mining and processing of uranium ore.' In addition to the ecological work, described in part above, the Research Institute has developed analytical chemistry, environmental radioactivity, geomorphological and toxicological research programs. Details of these programs are available in the Annual Research Reports from the Alligator Rivers Region Research Institute.

A major component of recent environmental investigation and assessment is concerned with predicting the effect of releasing mine-site water to the Magela Creek system. Total containment of all water within a restricted release zone in the project area has been required since the commencement of mining. The water management system at Ranger is designed to be in equilibrium with water demand for average climatic conditions. In years of higher rainfall, accumulation of water is likely unless an environmentally acceptable means of disposal is developed.

Table 6. Common and abundant birds of the Magela Creek system (Brennan, pers. comm.).

Pelecanidae	*Pelecanus conspicillatus*	Australian pelican
Anhingidae	*Anhinga melanogaster*	Darter
Phalacrocoracidae	*Phalacrocorax melanoleucas*	Little pied cormorant
	Phalacrocorax varius	Pied cormorant
	Phalacrocorax sulcirostris	Little black cormorant
Ardeidae	*Ardea pacifica*	Pacific heron
	Ardea picata	Pied heron
	Ardeola ibis	Cattle egret
	Egreta alba	Great egret
	Egreta garzetta	Little egret
	Egreta intermedia	Intermediate egret
	Nycticorax caledonicus	Rufous night heron
Ciconiidae	*Xenorhynchus asiaticus*	Jabiru
Plataleidae	*Plegadis falcinellus*	Glossy ibis
	Threskiornis aethiopica	Sacred ibis
	Threskiornis spinicollis	Straw-necked ibis
	Platalea regia	Royal spoonbill
Anatidae	*Anseranus semipalmata*	Magpie goose
	Dendrocygna arcuata	Wandering whistle duck
	Dendrocygna eytoni	Grass whistling duck
	Tadorna radjah	Radjah shelduck
	Anas superciliosa	Pacific black duck
	Anas gibberifrons	Grey teal
	Malacorhynchus membranaceus	Pink-eared duck
	Nettapus pulchellus	Green pygmy goose
Accipitridae	*Haliaeetus leucogaster*	White-bellied sea-eagle
Rallidae	*Porzana pusilla*	Baillon's crake
	Poliolimnas cinereus	White-browed crake
Gruidae	*Grus rubicundus*	Brolga
Jacanidae	*Irediparra gallinacea*	Comb-crested jacana
Charadriidae	*Vanellus miles*	Masked lapwing
	Erythrogonys cinctus	Red-kneed dotterel
	Charadrius melanops	Black-fronted plover
Recurvirostridae	*Himantopus himantopus*	Black-winged stilt
	Numenius minutus	Little curlew
	Tringa stagnatilis	Marsh sandpiper
	Calidris acuminata	Sharp-tailed sandpiper
	Calidris ferruginea	Curlew sandpiper
Glareolidae	*Stiltia isabella*	Australian pratincole
Laridae	*Chlidonias hybrida*	Whiskered tern
	Gelochelidon nilotica	Gull-billed tern

To assist in long-term water management on the mine-site, investigations have been carried out to develop receiving water quality standards for Magela Creek and to identify hazardous constituents in the ponded waters contained within the restricted release zone at the Ranger mine site. Standards were developed with respect to protection of both humans and the ecosystem. Risk to humans could occur from radiation exposure and metal intake. In deriving limits for the release of radionuclides the recommendations of the International Committee on Radiation Protection were followed. For non-carcinogenic metals, annual limits on intake were derived from National Health Medical Research Council recommendations on limiting concentrations in fresh food and from dietary details of critical groups. As carcinogenic risks are not well understood no limits on these risks have been included to date.

A major problem in setting standards for the protection of the ecosystem is the large number and complexity of species interrelationships with the water chemistry. In the absence of relevant experimental information a three stage process was adopted for deriving receiving water quality standards. The first stage was the adoption of a general criterion based on the observed natural fluctuations of the concentrations of the chemicals of concern in the natural creek water. This was followed by a hypothetical release from waterbodies on the mine-site into the Magela to identify specific constituents that would limit release volumes during the wet seasons. The criterion adopted was

"during a discharge period, the mean value of the altered distribution of each variable of concern is to be greater than the mean value of the natural distribution of the natural distribution, provided that a discharge mechanism is employed which ensures either a constant discharge rate or a proportional discharge rate of effluent to the creek."

The final stage was a toxicological and chemical assessment of the likely impact of increased concentrations of these limiting constituents on the creek ecosystem. The constituents limiting to water release from Ranger to Magela Creek following this procedure were sulphate and uranium, with magnesium and manganese imposing moderate limitation. An ecological and toxicological assessment on these elements indicated that for the latter two elements sufficient evidence existed to adopt values recommended by other environmental protection agencies. The toxicity of uranium to local aquatic animal species was assessed at the Alligator Rivers Region Research Institute (Allison et al. 1985) and, in addition an assessment was carried out on the effects of increased loads of sulphate, nitrate and phosphate on the ecosystem. A set of recommendations on interim receiving water quality standards was derived on the basis of this work.

Water has been released from areas on the mine-site outside of the restricted release zone; approximately 160,000 m^3 was released to Magela Creek during 1985. Effects observed in the mixing zone during the release included suppression of the reproductive capacity of the freshwater mussel (Velesunio angasi) and avoidance of released water by some fish species (Bishop and Walden 1985; Humphrey 1985). Further research is needed to determine whether these effects are environmentally significant and at what levels of dilution water release can be practiced.

As part of the consideration of the longer-term effects of water release from the mine-site the extent of recycling of nutrients and metals by floodplain plants has been investigated (Finlayson 1988). The floodplain ecosystem is dynamic and undergoes severe seasonal changes in water availability which affects the growth/decay cycle of many plants. It is postulated that some available nutrients and heavy metals are absorbed by the growing plants and released to the water/ sediment at the end of the growing season. Such recycling, and subsequent hydrologic dispersion of previously deposited substances could greatly affect the pattern of distribution and/or deposition of waterborne substances. It also raises the specter that secondary dispersion of these substances away from their initial deposition or absorption sites could occur.

The dominant grasses on the Magela Creek floodplain were sampled from Oct. 83 - Feb. 85 for changes in standing crop and chemical composition. By comparing the chemical composition of the grasses at the maximum and minimum standing crops an estimate of the amounts of nutrients and heavy metals that were initially accumulated and then released annually to the detrital component of the floodplain ecosystem can be obtained. A comparison of these amounts to the elemental loads contained in water storage ponds on the mine-site is presented in Table 7. For all elements considered the contents of the pond water does not exceed that contained in the grass detritus. Further information is being sought to determine the fate of these substances in the detritus and whether they are deposited elsewhere on the floodplain, incorporated into the sediments or exported to the East Alligator River. It is clear at this stage though, that the quantity of chemical elements biologically cycled and potentially mobilized by these grasses on the floodplain greatly exceeds that which would be added if all the water stored on the mine-site were released during one wet season.

In addition to cycling by the three grasses discussed other plants also contribute to the biological cycling and mobilization of chemical elements. In particular the extensive Melaleuca woodlands and forests are expected to play an important role. Steps are underway to determine the extent of this contribution, especially as early indications are that litter that falls during the dry season is physically removed and deposited elsewhere by wet season floods (Finlayson 1988).

During the 1983 late-dry season gamma-ray dose rates along transects of Magela Creek and floodplain were taken along with sediment cores and samples of P. spinescens in an attempt to determine the pattern of radionuclide accumulation in sediments and plants (Finlayson et al. 1985). The mean dose

Table 7. Elemental load from detritus from three aquatic grasses and in retention ponds at the Ranger mine site (adopted from Finlayson 1988).

	Estimate of total weight			
Element	Pseudoraphis	Oryza	Hymenachne	ponds
S (t)	110	10	200	91
Mg (t)	90	30	130	52
P (kg)	41 500	8 500	122 000	32
Mn (kg)	13 700	4 700	19 400	1531
Fe (kg)	123 600	186 200	130 000	1303
Cu (kg)	1 000	500	600	7
Zn (kg)	1 600	800	1 900	18
U (kg)	800	10	5	545
Pb (g)	4 100	32 300	8 800	4 600

rate on each transect plotted as a function of distance downstream of the mine demonstrated a peak in the Mudginberri Corridor where the creek spreads out in to the broad floodplain, the water velocity decreases, and dense beds of aquatic plants occur. From the end of the corridor to the outlet of the floodplain there is a monotonic increase in dose rates. Further analysis of samples is occurring in an attempt to determine whether specific areas on the floodplain act as sinks for particulate matter, including litter transported by the water. Initial indications suggest the perennial swamps along the western edge of the floodplain could be sinks.

The assessment of environmental impact by the Ranger Project is complicated by changes brought about by other activities in the area. The removal of large numbers of feral buffalo, Bubalus bubalis, from the Magela Creek floodplain is one such activity. Little quantitative evidence is available on the effect of this species on the floodplain ecology, but nevertheless the National Park Authority views its presence as incompatible with conservation objectives (see discussion on feral animals in Finlayson et al. 1988). The response of both native and alien species to the removal of pressures or competition formerly associated with buffalo is not known, and could make an assessment of the effect of the mine extremely difficult.

The recent introduction of the noxious weeds Mimosa pigra and Salvinia molesta (Finlayson 1984) is another event that might have significant environmental impact. While M. pigra has not spread very far, nor established thick stands, the floating S. molesta is now dispersed over the floodplain system, and at the end of the 1985 dry season completely covered three permanent billabongs. S. molesta and M. pigra, based on evidence elsewhere, are likely to have a serious effect on the floodplain biota. The presence of these weeds, like the feral buffalo, could be considered as imcompatible with the conservation aims of the Park Authority. The seriousness of these weed problems has been recognized, and effective manual control of M. pigra has occurred. Unfortunately, S. molesta has not been similarly controlled, despite introduction of biological control agents (Cowei et al. 1988).

The approach adopted in collecting ecological information on the Magela ecosystem has involved describing species and habitats, estimating abundance and seasonality, and in some instances, chemically analyzing biological material, sediments and water. It has also involved determination of the interrelationships and environmental parameters that sustain the ecosystem in its present state, and documenting the natural variability. Protection of the environment from effects caused by uranium mining involves prediction of the effects and assessment of their environmental significance. Experimental programs have been undertaken to determine what parameters can be modified, but still preserve the natural state, or what changes will occur as a result of modifying these parameters. The process of predicting the effect of proposed actions is made extremely difficult by the extent of variability of natural environmental conditions and ecological changes due to other activities in the area. As well as the effects of invasion and control of alien species the development of tourist resources, aboriginal utilization of the natural system and the conservation objectives of the National Park Authority have to be considered and assessed as part of the process of predicting the environmental consequences of mining uranium in the catchment of Magela Creek.

ACKNOWLEDGEMENTS

We are most grateful for the assistance and advice prodived by staff and consultants at the Alligator Rivers Region Research Institute. It has not been possible to relate the trials and tribulations associated with collecting the data that we have presented, but rest assured it was often an arduous undertaking and we thank all who contributed to the ecological sampling program. A special mention should be made of Jo Russell, our patient, cheerful and prompt typist.

REFERENCES

Allison, H., L. Baker, V. Brown, R. Byrne, S. Leighton and P. McBride. 1985. Effects of stream discharge: laboratory studies - toxicity of uranium to invertebrates. pp. 27-30 In Research Report 84-85, Supervising Scientist for the Alligator Rivers Region, Canberra, AGPS, Australia.

Bayly, I.A.E. and W.D. Williams. 1973. Inland waters and their ecology, Longman. Australia. 314 p.

Bishop, K.A. and M.A. Forbes. 1986. The freshwater fishes of northern Australia. In Haynes, C.D., Ridpath, M.A. and Williams, M.A.J., eds. Monsoonal Australia: Landscape, ecology and man in the northern lowlands, A.A. Balkema, Rotterdam, The Netherlands.

Bishop, K.A. and D. Walden. 1985. Effects of stream discharges: field studies - fish movement. pp. 14-17. In Research Report 84-85, Supervising Scientist for the Alligator Rivers Region, Canberra, AGPS, Australia.

Bishop, K.A., S.A. Allen, D.R. Pollard and M.J. Cook. 1981. Ecological studies on the fishes of the Alligator Rivers Region, Northern Territory. Supervising Scientist for the Alligator Rivers Region, Northern Territory, Australia, Open File Record 23.

Brady, H.T. 1979. Freshwater diatoms of the Northern Territory, especially in the Magela Creek system. In Brady, H.T., ed. The diatom flora of Australia Report 1, Macquarie University, Sydney, Australia.

Broady, P.A. 1984. A preliminary investigation of the Cyanobacteria ("blue-green algae") of the Magela Creek system and nearby sites, Northern Territory. Unpublished report to the Supervising Scientist for the Alligator Rivers Region, Northern Territory, Australia.

Brown, T.E., A.W. Morely and D.V. Koontz. 1985. The limnology of a naturally acidic tropical water system in Australia II. Dry season characteristics. Verhandlungen der Internationale Vereiningung Limnologie 22: 2131-2135.

Brown, T.E., A.W. Morley, N.T. Sanderson and R.D. Tait. 1983. Report of a large fish kill resulting from natural acid water conditions in Australia. Journal of Fisheries Biology 22: 335-350.

Christian, C.S. and M. Aldrick. 1977. A review report of Alligator Rivers Region environmental fact-finding study, Canberra, AGPS, Australia. 174 p.

Cook, C.D.K., B.J. Gut, E.M. Rix, J. Schneller and M. Seith. 1974. Water plants of the World, Dr. W. Junk. The Hague, The Netherlands.

Cowie, I.D., C.M. Finlayson and B.J. Bailey. 1988. Alien plants in the Alligator River Region, Northern Territory, Australia. Technical Memorandum 23, Supervising Scientist for the Alligator River Region, Northern Territory, Australia, Canberra, AGPS, Australia. 26 p.

Finlayson, C.M. 1984. Salvinia, water hyacinth and mimosa in the Alligator Region, Northern Territory. Australian Weeds 3: 83.

Finlayson, C.M. 1988. Productivity and nutrient dynamics of seasonally inundated floodplains of the Northern Territory. pp. 58-83. In D. Wade-Marshall and P. Loveday, eds. North Australia: Progress and Prospects. Vol. 2 Floodplain Research. ANU Press, Darwin, Australia.

Finlayson, C.M. and J.C. Gillies. 1982. Biological and physiochemical characteristics of the Ross River Dam, Townsville. Australian Journal of Marine and Freshwater Research 33: 811-827.

Finlayson, C.M., A. Johnston, A.S. Murray, R. Marten and P. Martin. 1985. Radionuclide distribution in sediments and macrophytes. Annual Research Report 84-85, pp. 88-90. In Supervising Scientist for the Alligator Rivers Region,Canberra, AGPS, Australia.

Finlayson, C.M., W.J. Freeland, B. Bailey and M. R. Fleming. 1988. Wetlands of the Northern Territory. pp. 103-126. In A.J. McComb and P.S. Lake, eds. The Australian Wetlands. Chipping North, Surrey-Beatty and Sons Pty. Limited, Australia. 196 p.

Finlayson, C.M., B.J. Bailey and I.D. Cowie, 1989.

Macrophyte vegetation of the Magela Creek floodplain, Alligator River Region, Northern Territory. Research Report 5, Supervising Scientist for the Alligator River Region, Northern Territory, Australia, Canberra, AGPS, Australia. 38 p.

Fox, R.W., C.G. Kelleher and C.B. Kerr. 1977. Ranger uranium environmental enquiry, second report, Canberra, AGPS, Australia.

Galloway, R.W. 1976. Summary description of the Alligator Rivers area. pp. 89-111. In Story, R., Galloway, R.W., McAlpine, J.R., Aldrick, J.M. and Williams, M.A.J., eds. Lands of the Alligator Rivers Region, Northern Territory, Land Research Series No. 38, Melbourne, CSIRO, Australia.

Grigg, G.C. and J.A. Taylor. 1980. An aerial survey of Crocodylus porosus nesting habitat in the Alligator Rivers Region, Northern Territory. Unpublished report to Australian National Parks and Wildlife Service, Canberra, Australia.

Humphrey, C. 1985. Effect of stream discharge: field studies - reproduction in freshwater mussels. pp. 18-19. In Annual Research Report 84-85, Supervising Scientist for the Alligator Rivers Region, Canberra, AGPS, Australia.

Jenkins, R.W.G. and M.A. Forbes. 1985. Seasonal variation in abundance and distribution of Crocodylus porosus in the tidal East Alligator River, Northern Territory. pp. 63-70. In Grigg, G., Shine, R. and Ehmann, H., eds. The biology of Australasian frogs and reptiles, Chipping North, Surrey Beatty and Sons, Australia.

Julli, M. 1985. The taxonomy and seasonal population dynamics of some Magela Creek floodplain microcrustaceans (Cladocera and Copepoda). Supervising Scientist for the Alligator Rivers Region, Northern Territory, Australia, Canberra, AGPS, Australia. 80 p.

Kessell, J.A. and P.A. Tyler. 1982. Phytoplankton populations of the Magela Creek system. Alligator Rivers Region, Northern Territory. Supervising Scientist for the Alligator Rivers Region, Northern Territory, Australia, Open File Record 18.

Legler, J.M. 1980. Taxonomy, distribution and ecology of freshwater turtles in the Alligator Rivers Region, Northern Territory. Supervising Scientist for the Alligator River Region, Northern Territory, Australia, Open File Record 2.

Ling, H.U. and P.A. Tyler. 1986. A limnological survey of the Alligator Rivers Region. II. Freshwater algae exclusive of diatoms. Research Report 3, Supervising Scientist for the Alligator Rivers Region, Northern Territory, Australia. Canberra, AGPS, Australia. 176 p.

McBride, T.P. 1983. Diatom Communities of the Muginberri Corridor, Northern Territory, Australia. Ph.D. thesis, Macquarie University, Sydney, Australia.

Marchant, R. 1982a. Seasonal variation in the macroinvertebrate faunas of billabongs along Magela Creek, Northern Territory. Australian Journal of Marine and Freshwater Research 33: 173-179.

Marchant, R. 1982b. Life spans of two species of tropical mayfly nymph (Ephemeroptera) from Magela Creek, Northern Territory. Australian Journal of Marine and Freshwater Research 33: 173-179.

Messel, H., A.G. Wells and W.J. Green. 1979. Surveys of tidal river systems in the Northern Territory of Australia and their crocodile population. The Alligator Region River system. Monograph No. 4, Pergamon Press, Sydney, Australia.

Morley, A.W. (ed.) 1981. Review of Jabiluka environmental studies, Sydney, Pancontinental Mining Pty. Ltd, Australia.

Morley, A.W., T.E. Brown and D.V. Koontz. 1985. The limnology of a naturally acidic tropical water system in Australia I. General Description and wet season characteristics. Verhandlungen der Internationale Vereinigung Limnologie 22: 2125-2130.

Morton, S.R. and K.G. Brennan. 1985. Birds. In Haynes, C.D., Ridpath, M.G. and Williams, M.A.J., eds. Monsoonal Australia: Landscape, ecology and man in the northern lowlands, A.A. Balkema, Rotterdam, The Netherlands.

Morton, S., K. Brennan and P. Dostine. 1984. Diversity, abundance and assessment of avian links in food chains. pp. 51-53. In Alligator Rivers Region Research Institute Research Report 1983-84, Canberra, AGPS, Australia.

Noller, B.N. and C. Hunt. 1985. Continuous measurement of physicochemical variables. pp. 68-69. In Alligator Rivers Region Research Institute Research Report 1984-1985, Canberra, AGPS, Australia.

Sadlier, R. 1981. A report on reptiles encountered in the Jabiru project area, Northern Territory. Supervising Scientist for the Alligator Rivers Region, Northern Territory, Australia, Open File Record 5.

Sanderson, N.T., D.V. Koontz and A.W. Morely. 1983. The ecology of the vegetation of the Magela Creek foodplain: Upper section from Oenpelli road crossing to Nankeen Billabong. Unpublished report in Scientific Workshop, Environment Protection in the Alligator Rivers Region, Jabiru, 17-20 May 1983, Australia.

Shine, R. 1986. Diets and abundances of aquatic and semi-aquatic reptiles in the Alligator River Region. Technical Memorandum 16, Supervising Scientist for the Alligator River Region, Northern Territory, Australia, Canberra, AGPS, Australia, 57 p.

Shine, R. and R. Lambeck. 1985. A radiotelemetric study of movements, thermoregulation and habitat utilization of Arafura file snakes (Serpentes: Acrochoididae). Herpetologica 41: 351-361.

Specht, R.L. 1958. The geographical relationships of the flora of Arnhem Land. pp. 415-478. In Specht, R.L. and Mountford, C.P., eds. Records of the American-Australian scientific expedition to Arnhem Land. 3. Botany and plant ecology, Melbourne University Press.

Story, R. 1976. Vegetation of the Alligator Rivers area, Northern Territory. pp. 89-111. In Story, R., Galloway, R.W., McAlpine, J.R., Aldrick, J.M. and Williams, M.A.J. eds. Lands of the Alligator Rivers area, Northern Territory, Land Research Series No. 38, Melbourne, CSIRO, Australia.

Tait, R.D., R.J. Shiel and W. Koste. 1984. Structure and dynamics of zooplankton communities, Alligator Rivers Region, N.T., Australia. Hydriobiologia 113: 1-13.

Taylor, J.A. 1979. The food and feeding habits of subadult Crocodylus porosus Schneider in northern Australia. Australian Wildlife Research 6: 347-359.

Taylor, J.A. and C.R. Dunlop. 1985. Plant communities of the wet-dry tropics of Australia: The Alligator Rivers Region. Proceedings of the Ecological Society of Australia 13: 83-103.

Thomas, D.P. 1983. A limnological survey of the Alligator Rivers Region I. Diatoms (Bacillariophyceae) of the region. Research Report 3, Supervising Scientist for the Alligator Rivers Region, Northern Territory, Australia. Canberra, AGPS, Australia.

Tyler, M.J. and M. Cappo. 1983. Diet and feeding habits of frogs of the Magela Creek system. Supervising Scientist for the Alligator Rivers Region, Northern Territory, Australia, Canberra, AGPS, Australia. 46 p.

Tyler, M.J., G.A. Crook and Davies M. 1983. Reproductive biology of the frogs of the Magela Creek system, Northern Territory. Records of the South Australian Museum 18: 415-445.

Vollenweider, R.A. 1968. Scientific fundamentals of the eutrophication of lakes and flowing waters with particular reference to nitrogen and phosphorus as factors in eutrophication. OECD, Paris, France.

Walker, T.D. and Tyler, P.A. 1982. Chemical characteristics and nutrient status of billabongs of the Alligator Rivers Region, Northern Territory. Supervising Scientist for the Alligator Rivers Region, Northern Territory, Australia, Open File Record 27.

Walker, T.D. and Tyler, P.A. 1983. Primary productivity of phytoplankton in billabongs of the Alligator Rivers Region, Northern Australia. Supervising Scientist for the Alligator Rivers Region, Northern Territory, Australia, Open File Record 8.

Walker, T.D. and P.A. Tyler. 1984. Tropical Australia, a dynamic limnological environment. Verhandlungen der Internationale Limnologie 22: 1727-1734.

Walker, T.D., J.T.O. Kirk and P.A. Tyler. 1982. The underwater light climate of billabongs of the Alligator Rivers Region, Northern Territory. Supervising Scientist for the Alligator Rivers Region, Northern Territory, Australia, Open File Record 20.

Walker, T.D., J. Waterhouse and P.A. Tyler. 1984. Thermal stratification and the distribution of dissolved oxygen in billabongs of the Alligator Rivers Region Northern Territory. Supervising Scientist for the Alligator Rivers Region, Northern Territory, Australia, Open File Record 28.

Webb, G. 1977. The natural history of Crocodylus porosus: Habitat and nesting. In Messel, H. ed. A study of Crocodylus porosus in northern Australia. A series of 5 lectures. Macarthur Press, Sydney, Australia.

Williams, A.R. 1979. Vegetation and stream pattern as indicators of water movement on the Magela floodplain, Northern Territory. Australian Journal of Ecology 4: 239-247.

ECOLOGY OF ATLANTIC WHITE CEDAR SWAMPS IN THE NEW JERSEY PINELANDS

Charles T. Roman[1]
Ralph E. Good

Division of Pinelands Research
Center for Coastal and Environmental Studies
Rutgers - The State University of New Jersey
New Brunswick, NJ 08903

and

Silas Little (retired)
Northeastern Forest Experiment Station
U.S. Forest Service

ABSTRACT

The areal extent of Atlantic White Cedar (Chamaecyparis thyoides) swamps in the 445,000 ha New Jersey Pinelands National Reserve has declined from over 20,000 ha in the late 19th century to less than 9,000 ha now occupying this coastal plain ecosystem. Much ecological research has been conducted on Pinelands cedar swamps with an ultimate objective of developing effective management programs aimed at maintaining and increasing the area of cedar. This paper reviews the literature that evaluates the vegetation response of Pinelands cedar swamps to disturbance factors, such as fire and cutting. A conceptual fire disturbance model is presented whereby several key variables interact to predict the potential for a burned stand to reestablish as cedar or convert to hardwoods (e.g., Acer rubrum).

[1]Current Address:
National Park Service
Coastal Research Center
University of Rhode Island
Narragansett, RI 02882

INTRODUCTION

Atlantic White Cedar (Chamaecyparis thyoides) swamps have been a focus of much scientific research in the New Jersey Pinelands. As stated by John Gifford (1900) "Nothing is more characteristic of the Coastal Plain of New Jersey than these swamps of cedar... Free from disease, and always a fresh rich green, cedar swamps form the most striking feature of the landscape." Besides being one of the more characteristic and aesthetic landscape features of the Pinelands they aid in maintaining the exceptional surface water quality which characterizes the region. The organic soils, dense vegetation and hydrologic regime of cedar swamps all contribute to a natural system that is effective at nutrient retention. Cedar swamps provide essential habitat for a diversity of plants and animals, including several rare species that reside and breed there (Wander 1981; Snyder and Vivian 1981). For example, curley grass fern (Schizaea pusilla), one of the more renowned endangered plants of the Pinelands, reaches its southernmost limit within the region's cedar swamps (Fairbrothers 1979). On a much broader scale, Pinelands cedar swamps provide a limited, yet valuable contribution toward enhancing landscape diversity within a region that is characterized by pitch pine (Pinus rigida) dominated communities (Forman 1979).

Most research on Pinelands cedar swamps has been related to forestry. Stone (1911) stressed the need for effective forest management programs aimed at insuring the long-term maintenance of Pinelands cedar resources. Beginning in the late 1800's, foresters (Gifford 1895; Pinchot 1900) began to report on silvical aspects of cedar in New Jersey. Later, Cottrell (1930) and Moore and Waldron (1940) investigated thinning to improve the quality of cedar swamps. The most comprehensive research was by Little (1950).

D. F. Whigham et al. (eds.), Wetland Ecology and Management: Case Studies, 163–173.

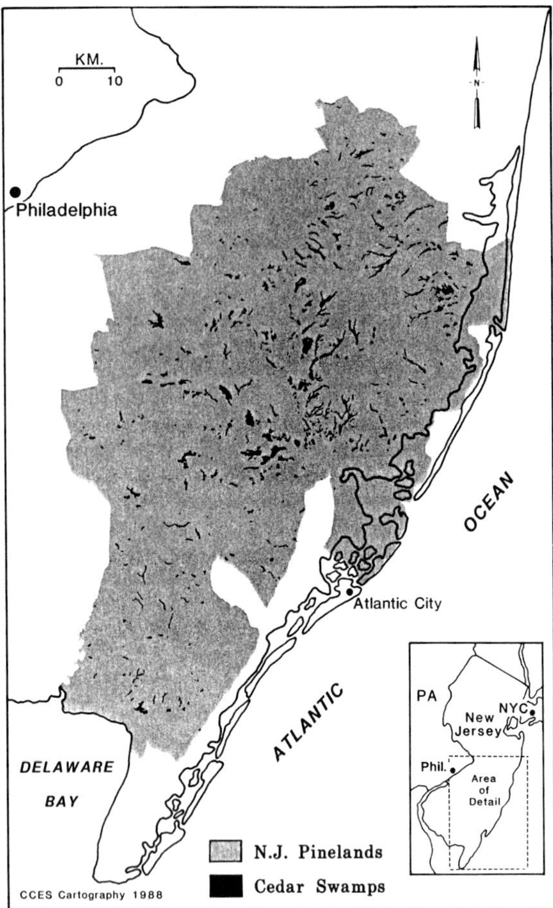

Figure 1. Distribution of Atlantic White Cedar (Chamaecyparis thyoides) swamps in the New Jersey Pinelands.

Little's (1950) monograph provides a foundation for this review on the ecology and management of Pinelands cedar swamps. The literature base that has accumulated over the past 35 years since that extensive monograph will be incorporated. Major topics discussed here will include the distribution, site characteristics and vegetation patterns and processes of Pinelands cedar swamps. Also addressed are management-protection strategies needed for the long-term perpetuation of cedar.

THE NEW JERSEY PINELANDS

The Pinelands, also known as the Pine Barrens, encompass about 445,000 ha, located almost entirely on the Outer Coastal Plain of New Jersey (Fig. 1). This area was never glaciated, although the Wisconsin Glaciation advanced to within approximately 30 km of the present Pinelands.

The region is underlain by sands and gravels ranging in thickness from 400 m at the western boundary to 1830 m at the Atlantic coast (Rhodehamel 1979a). The uppermost Cohansey Sand Formation is the principal aquifer or water table reservoir (Rhodehamel 1979b), a valuable natural resource of the area. Groundwater quality is generally exceptional, although the chemically inert, highly permeable substrate produces high potential for contamination (Means et al. 1981).

Dissecting the flat or gently undulating Pinelands topography are sluggish and relatively shallow streams flowing through the sand/gravel substrate (Patrick et al. 1979). Surface runoff is negligible and streams are fed predominantly by groundwater baseflow (Rhodehamel 1979b). Cedar swamps, hardwood swamps (e.g., red maple, Acer rubrum; blackgum, Nyssa sylvatica; sweetbay, Magnolia virginiana), and occasionally other wetland types border these streams. Stream water quality is characterized by high acidity and low pH (< 4.5), while nitrate-nitrogen levels near 0.05 mg l^{-1} are common.

The majority of upland and wetland areas are characterized by pitch pine, the Pinelands dominant tree species. Variable amounts of oak species (chiefly -- black oak, Quercus velutina; white oak, Q. alba; chestnut oak, Q. prinus; post oak, Q. stellata; blackjack oak, Q. marilandica) and other pines (shortleaf pine, P. echinata; Virginia pine, P. virginiana) are found in the canopy of the region's pine-oak and oak-pine upland forests. Two heaths, lowbush blueberry (Vaccinium vacillans) and black huckleberry (Gaylussacia baccata), along with scrub oak (Q. ilicifolia), dominate the upland forest understory.

Pinelands wetlands occupy about 35% of the region. They occur as complexes along stream courses or as isolated depressions, and include the predominant forested types such as cedar swamps, hardwood swamps and pitch pine lowlands. Some shrubs common to these forested wetlands include dangleberry (G. frondosa), highbush blueberry (V. corymbosum) and sweet-pepperbush (Clethra alnifolia), to name only a few. Shrub-thicket wetland communities dominated by leatherleaf (Chamaedaphne calyculata) and inland marshes are also encountered throughout the interior of the Pinelands.

NEW JERSEY PINELANDS CEDAR SWAMPS

Site Characteristics

Distribution

Cedar swamps occupy approximately

Table 1. Site characteristics of Pinelands cedar swamps. Depth of muck or peat above the sand substrate, water table depth relative to the forest floor, and surface pH are included.

Site Characteristic	Location	Investigator(s)
Muck/Peat Depth (m)		
0.6 to 0.9 (average) 3 (maximum)	Pinelands[1]	Waksman *et al*. 1943
1.2 to 2.1 (range)	Pinelands	Potzger 1945
1.5	McDonalds Branch	Buell 1970
2.5	Oswego River	Buell 1970
0 to 2 (range)	Pinelands	Givnish 1971
0.4 to 1.0 (average)	Pinelands	Markley 1979
Water Table Depth (cm)[2]		
-2.5 to 20 (average yearly range)	Pinelands	Little 1950
Surface (spring) 30 to 60 (summer)	Pinelands	Markley 1979
-17 to 17 (average yearly range)	Pinelands	Schneider and Ehrenfeld 1987[3]
pH		
4.0 (peat)[4]	Pinelands	Waksman *et al*. 1943
4.2 (peat)	Pinelands	Givnish 1971
3.4 to 4.8 (stream)	McDonalds Branch	Johnson 1979
4.5 (stream) 3.8 (peat)	Oyster Creek	Fusillo *et al*. 1981
2.8 (peat; undeveloped swamps)	Pinelands	Schneider and Ehrenfeld 1987[3]
4.45 (peat; developed swamps)		

[1]Data represent values from sites located throughout the Pinelands.

[2]Water table depth measured relative to the mean elevation of the forest floor. Negative denotes standing water above the forest floor.

[3]Schneider and Ehrenfeld present additional data not shown in this table for swamps in developed and undeveloped areas.

[4]pH data are presented for stream waters or standing water/near surface porewater from within the cedar swamp.

8,680 ha of the 445,000 ha New Jersey Pinelands (Roman and Good 1983). They are fragmented into 626 discrete (Fig. 1; Zampella 1987) patches. Nearly all (92%) of these patches are less than or equal to 40 ha. There are only four cedar swamps which are about 200 ha.

These cedar swamp patches are found bordering the slow flowing Pinelands streams from headwaters to areas under freshwater tidal influence. They generally occur as narrow bands along stream courses (< 300 m wide), yet some broad swamps do occur. Occasionally, an abrupt boundary between cedar swamp and upland forest will occur, but in most cases the cedar swamp is bordered or surrounded by hardwood swamp, pitch pine lowland or other wetland types.

Soils

The very poorly drained muck soil series supports most Pinelands cedar swamps. A typical soil profile consists of less than 1 m of highly organic muck, over a sand or gravelly sand substrate (Markley 1979). Although the muck layer (also referred to as peat) is generally shallow, Buell (1970) reports a depth of near 2.5 m in one Pinelands cedar swamp. Waksman et al. (1943) report that maximum peat depths in the Pinelands area are about 3 m (Table 1).

Hydrology and Water Quality

The water table of muck soils is generally at or above the forest floor during winter and spring and from 30-60 cm below the mean forest floor elevation in summer (Markley 1979). Little (1950) reports similar seasonal fluctuations (Table 1), while also stating that during long dry periods the water table may drop to as much as 90 cm below the mean forest floor in certain swamps.

Streams in the relatively undeveloped core area of the Pinelands, where approximately 60% of the region's cedar swamps occur, typically exhibit pH values of 4.5, or lower (Roman and Good 1983). The pH of standing water and near surface porewaters within cedar swamps is reportedly lower (Table 1). For instance, a three year data set at Oyster Creek (Ocean County, NJ) shows a median pH of 3.8 for waters on the swamp surface and a median value of 4.5 for the adjacent stream (Fusillo et al. 1980). Schneider and Ehrenfeld (1987), in an evaluation of several Pinelands cedar swamps, consistently found porewater pH values of less than 3.0.

Table 2. Some relatively common vascular plants of New Jersey Pinelands cedar swamps. For a more complete flora consult Stone (1911), Harshberger (1916), Little (1950, 1951), Givnish (1973) and Ehrenfeld and Schneider (1983). Nomenclature follows Fernald (1950).

Trees

Acer rubrum (Red Maple)
Chamaecyparis thyoides (Atlantic White Cedar)
Magnolia virginiana (Sweet Bay)
Nyssa sylvatica (Black Gum)
Pinus rigida (Pitch Pine)

Shrubs/Vines

Chamaedaphne calyculata (Leatherleaf)
Clethra alnifolia (Sweet Pepperbush)
Gaylussacia frondosa (Dangleberry)
Ilex glabra (Inkberry)
Ilex laevigata (Smooth Winterberry)
Kalmia angustifolia (Sheep Laurel)
Kalmia latifolia (Mountain Laurel)
Leucothoe racemosa (Fetterbush)
Lyonia ligustrina (Maleberry)
Rhododendron viscosum (Swamp Azalea)
Rhus vernix (Poison Sumac)
Smilax glauca (Sawbrier)
Smilax rotundifolia (Greenbrier)
Vaccinium atrococcum (Black Highbush Blueberry)
Vaccinium corymbosum (Highbush Blueberry)
Vaccinium macrocarpon (Cranberry)

Herbs

Carex spp. (Sedge)
Drosera rotundifolia (Sundew)
Orontium aquaticum (Golden Club)
Mitchella repens (Partridge-berry)
Sarracenia purpurea (Pitcher-Plant)
Utricularia spp. (Bladderwort)

Ferns

Osmunda cinnamomea (Cinnamon Fern)
Woodwardia aerolata (Netted Chain Fern)
Woodwardia virginica (Virginia Chain Fern)

Bryophytes and Lichens

Cetraria spp.
Cladonia spp.
Sphagnum spp.

Flora

Trees, shrubs, herbs, vines and ferns common to Pinelands cedar swamps are listed in Table 2. This list was developed based on a review of Stone's (1911) flora of southern New Jersey, Harshberger's (1916) description of the Pine Barrens cedar swamp formation, and more recent ecological studies of Pinelands cedar swamps (Little 1950, 1951; Schneider and Ehrenfeld 1987). More complete species lists can be found in these studies. For instance, Schneider and Ehrenfeld (1987) surveyed 19 individual cedar swamps within relatively undeveloped and developed watersheds throughout the Pinelands and list a total of 131 vascular plant species; however, the average number of species found at each site was much lower ranging from 27 species at undeveloped sites to 41 at developed sites. Species present at all sites included common trees (cedar, red maple, black gum), ericaceous shrubs (highbush blueberry, dangleberry, swamp azalea -- Rhododendron viscosum), other shrubs (sweet-pepperbush, winterberry -- Ilex laevigata), and some herbs (Sphagnum spp., cinnamon fern -- Osmunda cinnamomea).

Vegetation Patterns and Processes

Mature Swamp

Pinelands cedar swamp stands range from relatively mature to reproductive. These stands may be dominated by cedar or be mixed with hardwoods. The mature cedar swamps studied by Little (1950) ranged in age from 36-142 years, encompassing the age classes and stand characteristics of mature swamps evaluated by others (Table 3). McCormick's (1955, 1979) frequently cited description is of a mature swamp composed of a high density (basal area, 52 m^2 ha^{-1}) of relatively even-aged cedars forming an almost closed canopy (90% cover).

Table 3. Age, height and diameter at breast height (dbh) of mature cedar stands in the Pinelands.

Stand Characteristics			Location	Investigator(s)
Age (yrs)	Height (m)	dbh (cm)		
39-80[1]	7-20	6 - 28	Jobs Swamp	Harshberger 1916 Marigold Swamp
75	18	21	Shinns Branch	Moore and Waldron 1940[2]
36-142[1]	-	> 13	Pinelands	Little 1950
50	16	-	Middle Branch McDonalds Branch	McCormick 1955
-	12-18[1]	42 (maximum)	Cedar Bridge	Bernard 1963
70-130[1] 92	-	-	Shinns Branch	Hickman and Neuhauser 1977
85	-	28	Pinelands	Ehrenfeld and Schneider 1983[2]

[1]Range of canopy dominants is presented, while single numbers represent the mean of several measurements unless otherwise noted.

[2]Additional data are available in these reports.

Red maple, black gum and sweetbay form a continuous or relatively sparse tree understory. With openings in the canopy (e.g., created by windthrow or selective cutting) or along the swamp edges where light penetration is increased, these hardwoods often reach the canopy. Similarly, pitch pine may be scattered throughout the canopy. Some common shrubs often mixed with the understory trees or forming a lower stratum may include dangleberry, highbush blueberry, swamp azalea, sweet pepperbush, fetterbush (Leucothoe racemosa) and bayberry (Myrica pensylvanica). The herbaceous layer of McCormick's (1955) mature swamp (aside from the conspicuous Sphagnum spp.) is sparse (4% cover); however, species diversity is particularly high (Harshberger 1916; Little 1951; McCormick 1955, 1979). Chain ferns (Woodwardia aerolata, W. virginica), sundews (Drosera spp.), bladderworts (Utricularia spp.) and pitcher plants (Sarracenia purpurea) are fairly common.

Vegetation Development and Disturbance

Little (1950) suggests that the mature Pinelands cedar swamp, if protected from natural or human-induced disturbances such as fire or cutting, will ultimately develop into a hardwood swamp. Red maple would probably dominate, with black gum and sweetbay as principal associates. The shrub layer would be more dense and continuous than in the cedar swamp and dominated by sweet-pepperbush and highbush blueberry. This vegetation change would occur very gradually as the veteran cedar trees died, with replacement by understory hardwoods, thereby grading through the mixed cedar-hardwood type.

This successional hypothesis is partly related to the different responses of cedar and hardwood seedlings to low light levels. Cedar seedlings are initially successful under the dense canopy of a mature cedar stand, but they only survive from 1-3 years (Little 1950). Hardwood seedlings (red maple, blackgum, sweetbay) also suffer high mortality under low light conditions, however, being more shade tolerant than cedar, some seedlings survive for longer than 3 years (Little 1950). A relatively continuous size and age class distribution of hardwoods results, often ranging from seedling/sapling to understory status. With gradual dying of mature cedars, the hardwood understory is greatly favored for release to the canopy.

In the Pinelands, wildfire, cutting, storm damage, flooding/draining, or cranberry cultivation almost inevitably interrupt predictable succession. For example, Forman and Boerner (1981) determined that an average point in the Pinelands burns at about 65 year intervals. While this fire frequency is not specific to cedar swamps, it stresses the important role that fire may play in shaping vegetation development in Pinelands cedar swamps. Cedar swamps that remain protected from wildfire may be disturbed by other forces. Wind speeds of 96-112 km h^{-1} occur in the Pinelands at a 10 year frequency (Thom 1963). Addess and Shulman (1966) report an average of four cases per year of freezing rain in the Pinelands. These meteorological events have the potential to create canopy gaps, thus altering vegetation development. Cutting represents the major human- induced disturbance to Pinelands cedar swamps. Extensive harvesting of cedar began in the 1700's, and by 1749 Peter Kalm was warning of overexploitation in the Pinelands (Wacker 1979).

Fire Disturbance

To introduce the many variables that interact to shape the vegetation development of Pinelands cedar swamps, fire disturbance will be addressed. Cedar swamps and associated wetland complexes often serve as firebreaks. However, severe fires may burn into cedar swamps. Wetland complexes of less than 300 m wide are almost always crossed by wildfires that approach perpendicular to the wetland margins (Windisch 1987). When the water table is above or near the surface, the fire may rapidly burn through the crowns, leaving the forest floor unburned or scorched at most (Little 1979). In contrast, during dry periods the organic matter may burn to the water table or underlying mineral substrate. Regardless of the fire intensity, cedar trees with thin bark are usually killed and do not sprout (Korstian and Brush 1931).

Vegetation composition of stands following fire is primarily determined by the depth of the forest floor consumed, the stand composition prior to fire, and available seed sources (Little 1950, 1979). These, and other interrelated factors are shown in a conceptual cedar swamp - fire disturbance model (Fig. 2). If significant quantities of organic matter were consumed, an area of standing water may be created. Such an area would remain open, be colonized by Sphagnum spp., be invaded by a shrub such as leatherleaf, or develop a mosaic of wetland types. Establishment of tree seedlings would probably be limited by the high water table (except on hummocks) and dense cover of leatherleaf. However, with a gradual accumulation of organic matter and Sphagnum spp. a suitable bed for seedling establishment could develop. Species of seed from nearby

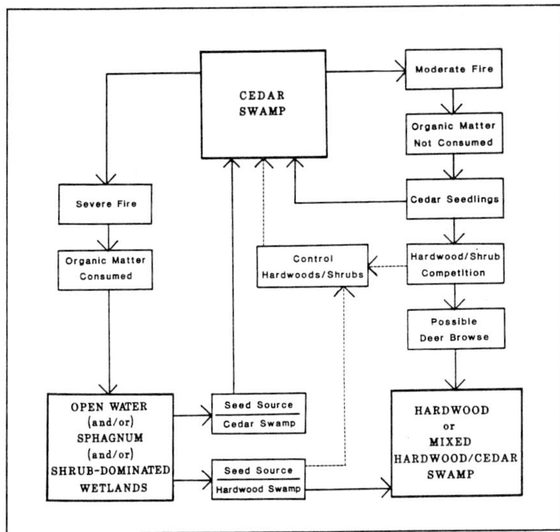

Figure 2. Conceptual cedar swamp-fire disturbance model.

sources would determine the ultimate forest composition.

Following a fire that does not consume the forest floor, cedar may reproduce from seed stored in the peat (Fig. 2). Under a mature cedar stand, Little (1950) found that the top 2.5 cm of forest floor could contain 640,000-2,700,000 viable seeds per ha. With this profusion of seedlings, which could be augmented by seed input from an adjacent unburned stand, it seems that a relatively pure cedar stand could develop, especially when considering the optimal full light condition. However, other variables interact. For example, if hardwoods and shrubs were prevalent in the understory prior to the fire, then sprouts may out-compete the cedar with ultimate development of a hardwood or mixed stand. Browsing of cedar seedlings by deer represents a further interacting variable that generally favors the establishment of hardwood or mixed stands. After burning of a mature cedar stand, Little and Somes (1965) established a fenced deer exclusion plot and an unfenced plot. Cedar reproduction was initially high in these plots. Nine years later both the density and height of surviving seedlings in the unfenced (browsed) plot were so low that ultimate regeneration to cedar would be either unlikely or delayed.

To summarize the fire disturbance scenario (Fig. 2), several variables (depth of the forest floor consumed, stand composition, seed source, browsing) have been cited as interacting to shape the vegetation development of burned cedar stands. Little (1950, 1979) notes additional variables, most notably water table level. This interaction of variables has generally favored the establishment of hardwoods with a corresponding decrease in cedar swamps (Little 1950, 1979). However, this should not imply that pure cedar stands never establish following fire. Reestablishment of cedar from seeds stored in the peat would be favored if the hardwood understory were only sparsely developed prior to the fire, and especially if local deer populations were low. If a fire burned the peat so that the possibility of hardwood re-sprouting were eliminated, cedar could potentially become established by input of wind dispersed seed from an adjacent unburned stand.

Disturbance Summary

The general trend in response to disturbance, especially fire and cutting, has favored conversion of cedar swamps to hardwood or mixed hardwood-cedar swamps. However, with the proper suite of interacting factors cedar swamps do perpetuate following disturbance. In fact, disturbance is required for maintenance of cedar in the Pinelands. As previously discussed, Little (1950) suggests that without disturbance the aging cedar swamp would be replaced by hardwoods. Preceeding European colonization, natural disturbances such as windthrow, beaver activity, and fire may have contributed to the reestablishment of cedar in areas where a mature stand, of 100 years old or more, was declining (Little 1950).

Today, these same disturbances, along with poorly managed logging activities, may actually hasten vegetation development to hardwoods. This may be a function of a lower fire frequency than what occurred in the past, as suggested by Forman and Boerner (1981). Or, deer browsing may be more concentrated in cedar swamps than in the past, because of higher populations or loss of habitat and other food sources. Also, swamp hardwoods may be more abundant in today's forest, thereby enhancing their competitive potential. To conclude, it is recognized that disturbance is necessary and has been responsible for the perpetuation of cedar swamps in the Pinelands. However, considering recent trends in conversion of cedar to hardwood swamps and other wetland types, it seems that disturbance must be carefully managed or controlled in order to encourage, not deter, cedar regeneration.

Status and Management

Areal Estimates

The decline in cedar swamps of the Pinelands is supported by historical accounts

Table 4. Estimates of cedar swamp area in the New Jersey Pinelands.

Year	Area (ha)	Method	Investigator(s)
circa 1899	21,250	Field Reconnaissance	Vermuele 1900
circa 1930	40,500	Field Reconnaissance	Cottrell 1930
1956	12,050	Aerial photo interpretation/vegetation map	McCormick and Jones 1973
1972	20,150	USFS Forest Survey	Ferguson and Mayer 1974
1979	8,680	Aerial photo interpretation/vegetation map	Roman and Good 1983; Andropogon Associates 1980

(Wacker 1979) and by areal estimates (Table 4). Although comparing areal estimates by different investigators must be done with caution, an apparent decrease is noted when comparing the 1899, 1956, and 1979 values. Cottrell (1930) implies that mixed stands are included in his 1930 estimate, possibly contributing to the seemingly high area. The 1972 estimate (Ferguson and Mayer 1974) may also be high for the same reason, but moreover, these authors report a substantial sampling error. The most meaningful comparison is made between the 1956 and 1979 estimates, derived from aerial photographs interpreted by the same individuals. During this 23 year period there has been a loss of over 3000 ha, or 1/3 of the total cedar area estimated to be present in 1956. Although possibly coincidental, Ferguson and Mayer (1974) also note a decrease of 1/3 between an initial 1956 US Forest Service survey (Webster and Stolenberg 1958) and the 1972 survey.

Cedar Swamp Management

The perpetuation of cedar swamps requires an effective management program. The careful regulation of cedar harvesting seems to have the greatest potential for implementation. All cedar harvests in the Pinelands must comply with standards set forth in the Forestry Management Program of the Pinelands Comprehensive Management Plan (New Jersey Pinelands Commission 1980). To favor reestablishment of cedar in harvested areas, harvesting must be by clearcutting, slash must be reduced or removed and competing hardwoods must be controlled (Fig. 3). In addition, Zampella's (1987) conceptual harvest management model establishes the basis for

Figure 3. Recent harvest of a portion of Atlantic White Cedar (Chamaecyparis thyoides) swamp in the New Jersey Pinelands. Note windthrow trees at the margins of the harvest area and poor cedar regeneration.

controlled disturbances to encourage the maintenance of cedar in the Pinelands. According to Zampella (1987), prior to cutting, a site evaluation should be conducted to predict the probability of successful cedar reestablishment. Factors to consider would include swamp size, shape and orientation, stand age, condition and composition, hydrologic conditions, adjacent forest type, and local deer population. From such an analysis harvests could be directed to more favorable sites.

In addition to harvest management, a complete resource management program for cedar swamps must include land acquisition. Nearly 50% of Pinelands cedar swamps are on state-owned lands (Zampella 1987). On these and other conservation lands, management should be directed toward maintaining and increasing the area in cedar. Existing stands should be grown on a relatively long rotation, often with cutting delayed until the cedar canopy begins to break up through windthrow or other causes. Trees in younger stands killed by fire or flooding would be salvaged if merchantable. When labor and suitable techniques are available, conversion of many mixed or hardwood stands to cedar would be desirable. It is likely that such management directed toward maintaining and increasing the area in cedar would be most effective on publicly owned lands where proper planning (e.g., initial site evaluation) and long-term management (e.g., slash reduction and hardwood control) could be readily implemented.

Fire has been identified as a factor that often converts cedar to hardwood swamps. This trend could change if competing hardwoods were controlled (Fig. 2). As with maintenance harvest, management of burned swamps could be effectively implemented on publicly-owned conservation lands. Following the salvage harvest, hardwoods should be controlled by selective cutting or selective herbicide use on sites determined to have a high potential for reestablishment as cedar. Such control or management should enhance the probability of cedar reestablishment.

CONCLUSION

Long-term management strategies

The perpetuation of cedar swamps in the Pinelands is mostly contingent upon effective implementation of 1) cedar swamp management and 2) prohibiting development of swamps and adjacent lands. The resource management program should include a) a harvest management model aimed at predicting the probability of successful cedar reestablishment following harvest of a particular stand, b) effective post-harvest control of slash and competing hardwoods, and c) dedicated efforts toward maintaining and increasing the area of cedar swamps within state-owned and other conservation lands.

With respect to regulatory protection of cedar swamps, the Pineland's Wetlands Management Program prohibits most development on wetlands (New Jersey Pinelands Commission 1980; Good and Good 1984; Zampella and Roman 1983), and requires that an upland buffer area of at least 91 m (300 ft) must be maintained, unless the applicant can demonstrate that a lesser buffer would not impact the wetland. To aid in implementation of this buffer requirement, Roman and Good (1986) developed a model for delineating the minimum site-specific buffer width required to protect a wetland from potential development impacts. For cedar swamps, this model recommends no exemptions to the 91 m requirement.

Strategies for the successful long-term maintenance of Pinelands cedar swamps are currently in place. However, these various programs and models must be strictly enforced and monitored to insure that cedar swamps remain as both ecologically and economically valuable assets of the Pinelands ecosystem.

ACKNOWLEDGEMENTS

We thank George H. Pierson (New Jersey Bureau of Forest Management) and Robert A. Zampella (New Jersey Pinelands Commission) for their review of the manuscript and many helpful suggestions.

REFERENCES

Addess, S.R. and Shulman, M. D. 1966. A synoptic analysis of freezing precipitation in New Jersey. Bulletin of the New Jersey Academy of Science 11: 2-7.

Andropogon Associates. 1980. Forest vegetation of the Pinelands. New Jersey Pinelands Commission, New Lisbon, NJ. USA. 42 p.

Bernard, J.M. 1963. Lowland forests of the Cape May formation in southern New Jersey. Bulletin of the New Jersey Academy of Science 8:1-12.

Buell, M.F. 1970. Time of origin of New Jersey Pine Barrens bogs. Bulletin of the Torrey Botanical Club 97: 105-108.

Cottrell, A.T. 1930. Thinning white cedar in New Jersey. Journal of Forestry 28:1157-1162.

Ehrenfeld, J.G. and Schneider., J.P. 1983. The sensitivity of cedar swamps to the effects of non-point source pollution associated with suburbanization in the New Jersey Pine Barrens. Technical Report. Center for Coastal and Environmental Studies, Rutgers-The State University of NJ, New Brunswick, NJ. USA. 42 p.

Fairbrothers, D.E. 1979. Endangered, threatened, and rare vascular plants of the Pine Barrens and their biogeography. pp. 395-405 In Forman, R.T.T, ed. Pine Barrens: ecosystem and landscape. Academic Press, New York, NY. USA.

Ferguson, R.H. and Mayer. C.E. 1974. The timber resources of New Jersey. Northeastern Forest Experiment Station, Forest Service, U.S. Dept. of Agriculture, Upper Darby, PA. USDA Forest Service Resource Bulletin NE-34. 58 p.

Fernald, M.L. 1950. Gray's manual of botany. American Book Co., New York, NY. USA. 1632 p.

Forman, R.T.T. (ed.). 1979. Pine Barrens: ecosystem and landscape. Academic Press, New York, NY, USA. 601 p.

Forman, R.T.T. and Boerner, R.E. 1981. Fire frequency and the Pine Barrens of New Jersey. Bulletin of the Torrey Botanical Club 108:34-50.

Fusillo, T.V., Schornick, R.E. Jr., Koester, H.E. and Harriman, D.E. 1980. Investigation of acidity and other water-quality characteristics of Upper Oyster Creek, Ocean County, New Jersey. U.S. Geological Survey, Water Resources Division, Trenton, NJ. Water Resources Investigations 80-10. 28 p.

Gifford, J. 1895. A preliminary report on the forest conditions of South Jersey. pp. 245-286 In Annual report of the state geologist for 1894. Trenton. NJ. USA.

Gifford, J. 1900. Forestal conditions and silvicultural prospects of the coastal plain of New Jersey. pp. 235-327. In Annual report of the state geologist for the year 1899, report on forests. Trenton, NJ. USA.

Givnish, T.J. 1971. A study of New Jersey Pine Barrens cedar swamps. Technical Report. Princeton-NSF Cedar Swamp Study Group, Princeton, NJ. USA.

Good, R.E. and Good, N.F. 1984. The Pinelands National Reserve: an ecosystem approach to management. BioScience 34:169-176.

Harshberger, J.W. 1916. The vegetation of the New Jersey Pine Barrens. Christopher Sower Co., Philadelphia, PA. Reprinted, Dover Publ., New York, NY. USA. 329 p.

Hickman, J.C. and Neuhauser, J.A. 1977-78. Growth patterns and relative distribution of Chamaecyparis thyoides and Acer rubrum in Lebanon State Forest, New Jersey. Bartonia 45:30-36.

Johnson, A.H. 1979. Acidification of headwater streams in the New Jersey Pine Barrens. Journal of Environmental Quality 8:383-386.

Korstian, C.F. and Brush, W.D. 1931. Southern White Cedar. U.S. Dept. of Agriculture, Washington, DC. Technical Bulletin 251:1-76.

Little, S. 1950. Ecology and silviculture of white cedar and associated hardwoods in southern New Jersey. Yale University School of Forestry Bulletin 56:1-103.

Little, S. 1951. Observations on the minor vegetation of the Pine Barrens swamps in southern New Jersey. Bulletin of the Torrey Botanical Club 78:153-160.

Little, S. 1979. Fire and plant succession in the New Jersey Pine Barrens. pp. 297-314. In Forman, R.T.T., ed. Pine Barrens: ecosystem and landscape. Academic Press, New York, NY. USA.

Little, S. and Somes, H. 1965. Atlantic white cedar being eliminated by excessive animal damage in South Jersey. U.S. Forest Service Research Note NE-33. Upper Darby, PA. USA. 3 p.

Markley, M.L. 1979. Soil series of the Pine Barrens. pp. 81-93. In Forman, R.T.T. ed. Pine Barrens: ecosystem and landscape. Academic Press, New York, NY. USA.

McCormick, J.S. 1955. A vegetation inventory of two watersheds in the New Jersey Pine Barrens. PH. D. Dissertation. Rutgers-The State University of NJ, New Brunswick, NJ. USA. 125 p.

McCormick, J. 1979. The vegetation of the New Jersey Pine Barrens. pp. 229-243. In Forman, R.T.T. ed. Pine Barrens: ecosystem and landscape. Academic Press, New York, NY. USA.

McCormick, J. and Jones, L. 1973. The Pine Barrens: vegetation geography. NJ State Museum, Research Report No. 3. Trenton, NJ. USA.

Means, J.L. Yureteich, R.F., Crerar, D.A., Kinsman, D.J. and Borcsik, M.P. 1981. Hydrogeochemistry of the New Jersey Pine Barrens. NJ Dept. of Environmental Protection and NJ Geological Survey, Trenton, NJ. USA. 107 p.

Moore, E.B. and Waldron, A.T. 1940. Growth studies of southern white cedar in New Jersey. Journal of Forestry 38:568-572.

New Jersey Pinelands Commission. 1980. Comprehensive management plan for the Pinelands National Reserve (National Parks and Recreation Act, 1978) and Pinelands Area (Pinelands Protection Act, 1979). New Lisbon, NJ. USA. 446 p.

Patrick, R., Matson, B. and Anderson, L. 1979. Streams and lakes in the Pine Barrens. pp. 169-193. In Forman, R.T.T., ed. Pine Barrens: ecosystem and landscape. Academic Press, New York, NY. USA.

Pinchot, G. 1900. Silvicultural notes on the white cedar. pp. 131-135. In Annual report of the state geologist for the year 1899, report on forests. Trenton, NJ. USA.

Potzger, J.E. 1945. The Pine Barrens of New Jersey, a refugium during Pleistocene times. Butler University Botanical Studies 7:1-15.

Rhodehamel, E.C. 1979a. Geology of the Pine Barrens of New Jersey. pp. 39-60. In Forman, R.T.T., ed. Pine Barrens: ecosystem and landscape. Academic Press, New York, NY. USA.

Rhodehamel, E.C. 1979b. Hydrology of the New Jersey Pine Barrens. pp. 147-167. In Forman, R.T.T., ed. Pine Barrens: ecosystem and landscape. Academic Press, New York, NY. USA.

Roman, C.T. and Good, R.E. 1983. Wetlands of the New Jersey Pinelands: values, functions, and impacts. Technical Report. Division of Pinelands Research, Center for Coastal and Environmental Studies, Rutgers-The State Univ. of NJ, New Brunswick, NJ. USA. (revised 1986). 82 p.

Roman, C.T. and Good, R.E. 1986. Delineating wetland buffer protection areas: the New Jersey Pinelands model. pp. 224-230. In Kusler, J.A. and Riexinger, P., eds. Proceedings of the National Wetland Assessment Symposium. Association of State Wetland Managers, Chester, VT. USA.

Schneider, J.P. and Ehrenfeld, J.E. 1987. Suburban development and cedar swamps: effects on water quality, water quantity, and plant community composition. pp. 271-288. In Laderman, A.D., ed. Atlantic White Cedar wetlands. Westview Press, Boulder, CO. USA.

Snyder, D. and Vivian, V.E. 1981. Rare and endangered vascular plant species of New Jersey. Conservation and Environmental Studies Center and U.S. Fish and Wildlife Service. 98 p.

Stone, W. 1911. The plants of southern New Jersey with especial reference to the flora of the Pine Barrens and the geographic distribution of the species. NJ State Museum, Annual Report, Trenton, NJ. USA. pp. 21-828,

Thom, H.C. 1963. Tornado probabilities. Monthly Weather Review 9:730-748.

Vermeule, C.C. 1900. The forests of New Jersey. pp. 13-108. In Annual report of the state geologist for the year 1899, report on forests. Trenton, NJ. USA.

Wacker, P.O. 1979. Human exploitation of the New Jersey Pine Barrens before 1900. pp. 3-23. In Forman, R.T.T., ed. Pine Barrens: ecosystem and landscape. Academic Press, New York, NY. USA

Waksman, S.A., Schulhoff, S., Hickman, C.A., Cordon, T.C. and Stevens, S.C. 1943. The peats of New Jersey and their utilization. NJ Department of Conservation and Development, Geological Series Bulletin 55 (Part B). Trenton, NJ. USA. 278 p.

Wander, W. 1981. Breeding birds of southern New Jersey cedar swamps; occasional paper No. 138. New Jersey Audubon (Records of N.J. Birds) 6:51-65.

Webster, H.H. and Stoltenberg, C.R. 1958. The timber resources of New Jersey. Northeastern Forest Experiment Station, Forest Service, U.S. Department of Agriculture, Upper Darby, PA. USA. 41 p.

Windisch, A.G. 1987. The role of stream lowlands as firebreaks in the New Jersey Pine Plains region. pp. 313-316. In Laderman, A.D., ed. Atlantic White Cedar Wetlands. Westview Press, Boulder, CO. USA.

Zampella, R.A. 1987. Atlantic white cedar management in the New Jersey Pinelands. pp. 295-311. In Laderman, A. D., ed. Atlantic White Cedar Wetlands. Westview Press, Boulder, CO. USA.

Zampella, R.A. and Roman, C.T. 1983. Wetlands protection in the New Jersey Pinelands. Wetlands 3:124-133.

SUBJECT INDEX

S

T

U

V

W

X

Z